Introduction to Facilities, Utilities & Cleanroom Qualification for Pharmaceuticals and Biologics

Ben O'Shea

ISBN: 9798321090954
Imprint: Independently published

© First edition, 2024
Modular MedTech Series

Disclaimer: This publication is intended for general learning and general training purposes. While every effort has been made by the author to ensure informative and accurate information is provided in this publication, the author and publisher shall not be responsible for any error or omission, and or subsequent material loss, financial, personal, private, public, property or physical damage or harm. It is the responsibility of individuals to conform with all legal and regulatory requirements in regard to the topics covered in this publication, and to apply the standards, practices and applicable regulations to the applications, products, business and commercial activities they are involved with.

Contents

1. Introduction to Commissioning, Qualification And Validation 8
1.1. Overview 8
1.2. Regulatory Requirements 9
2. EU GMP V4 Annex 15 23
2.1. Introduction 23
2.2. Changes to validated systems or processes 23
2.3. Validation and product lifecycle 23
3. Cleanroom Qualification 24
3.1. INTRODUCTION 24
3.2. Cleanroom Environment 25
3.3. Cleanroom Zoning and Classification 26
3.4. Types of Contamination 27
3.5. Cleanroom Classification Table 27
3.6. Zone Classification 29
3.7. HVAC Particulate Control 29
3.8. Total Airflow Volumes & Recovery Rates 29
3.9 Unidirectional flow 31
3.10. Temperature 39
3.11. Relative Humidity (RH) 39
3.12. Control and Spread of Smoke 39
3.13. Cleaning 39
4.1 Introduction 40
4.2. HVAC System Design 40
4.3. ISO Standards for Cleanrooms 44
4.4. Temperature 44
4.5. Air Handling Units 44

4.6. Filtration 44

4.7. course/ PRE- Filtration 47

4.8. Fine / SECONDARY Filtration 47

4.9. Compliance Tests for GMP Zones 47

4.10. Clean Room Design Considerations 48

5. Utility Gases & Water 51

5.1. INTRODUCTION 51

5.2. CLEAN STEAM 51

5.3. RO Water, DI water and Water for Injection 53

5.3.1. Water Systems 54

5.4. High Purity Water System Design 54

5.5. Water for Injection 56

WFI generation, storage and distribution 56

5.6. Microbial Limits 59

5.7. Purified Water Systems 59

6. Compressed Air- Generation, storage and distribution 60

6.1. Introduction 60

6.2. Compressed Air Design Requirements 61

6.3. Design Element: Inlet Air filters 61

6.4. Design Requirements 62

6.5 Design Qualification 63

6.6. DQ Evaluation 63

7. Clean steam 65

7.1. Introduction 65

8. Facilities Monitoring 67

8.1. OVERVIEW 67

9. Gase Systems 69

9.1. Argon Gas 69

9.2. CARBON DIXOIDE 69

9.3. oxygen 70

10. Steam Sterilization 72

10.1. Sterilization and Disinfection 72

10.2. Parametric Approach 72

10.3. Risk and Sterility 74

10.4. Spaulding's classification 74

10.5. Cleaning 74

10.6. Clean-in-Place (CIP) 75

10.6.1. PIC/S Guidance on Limits 76

10.7. Antimicrobial Techniques 76

10.7.1. Pasteurization 76

10.8. Sterilisation Processes 77

10.9. FDA Categorisation of Established Sterilization Processes 77

10.10. Steam Sterilizer (Moist Heat) - Development of Sterilization Processes 79

10.11. The Sterilizer as Equipment 81

10.12. The Sterilization Process 82

10.13. Validation of Steam Sterilizers 84

10.14. Requalification 86

10.15. Industry Standards relevant to Sterilization 86

10.16. Principle of Operation 89

11. Alternatives to Steam Sterilization 91

11.1. Ethylene Oxide (EO) 91

11.2. Oxidizing and Non Oxidizing Disinfectants 92

11.3. Sodium hypochlorite 92

12. Depyrogenation 93

12.1. Pyrogens 93

12.2. Bacterial Toxins 93

12.3. Pyrogen Assay - Limulus Amoebocyte Lysate 94

12.4. Endotoxins and Depyrogenation 94

12.5. Biological Indicators for Dry Heat 95

12.6. Control of Materials 95

12.7. Contamination Considerations 95

12.8. Start-up Conditions..

12.9. In-Process Controls 96

12.10. Cooling 96

12.11. Failure of Depyrogenation 96

12.12. Depyrogenation -Performance Qualification (PQ) 97

13. Aseptic Processing 100

13.3. Design Considerations for Isolator Systems 101

13.4. Definition of Aseptic Processing 103

13.5. Regulations and Standards 103

13.6. Technical Comparison of Terminal Sterilization and Aseptic Processing 104

13.7. Isolator and Glove Access 105

13.8. Isolator Design Requirements 105

13.9. Materials of Construction 105

13.10. Isolator Access 106

13.11. Isolator Decontamination 106

13.12. Isolator Barrier Systems 109

13.13 Isolator Interfaces 111

13.14. Isolator Decontamination 111

13.15. Facility Layout for Aseptic Processing 112

13.16. Air Classifications 113

13.17. Filling Operations 113
13.18. Aseptic Process Simulation 114
14. Sterile Barrier Packaging Systems 115
14.1. Introduction 115
14.2. Material Compatability 118
14.3. Facators in Design and Development 118
14.4. Performance of Packaging System 118
14.5. Stability of Packaging 119
14.6. Lifecycle approach to Sterile Barrier Systems 120
14.7. Factors in Sterile Barrier Validation 120

1. Introduction to Commissioning, Qualification And Validation

1.1. Overview

The Qualification of facilities and utilities is best managed with the creation of a qualification plan. The plan can provide a framework that outlines the qualification activities, rationales, deliverables, resources and timing. However, certain qualification activities are strongly recommended and mandated by health regulators especially within pharmaceutical biotech, MedTech and medical device sectors.

The regulatory legislation pertaining to the specific products and markets can inform the essential qualification requirements. Medical devices range in their principal mechanism of action, complexity and intended use. For example, the facility and supporting utilities necessary for the manufacture and packing of a surgical implant differs from an Orthopedic crutch or aid. Yet again, a medicinal product or combination device such as a pre-filled syringe with a biological formulation will require aseptic techniques to be applied during the process. This controlled environment that assures sterility is supported by qualified facilities and utilities that need to function and perform consistently.

Therefore, the scope and complexity of C&Q and validations must be designed based on the products manufactured and their intended purposes. With that said, there are a number of keystone commissioning, qualification and validation activities that represent best practices that are broadly applied to meet regulations. The essentials of C&Q can be specified in company (in-house) procedures or standard operating procedures. The discrete requirements required for specific projects can then be guided with the creation of a C&Q plan.

High Level understanding C&Q

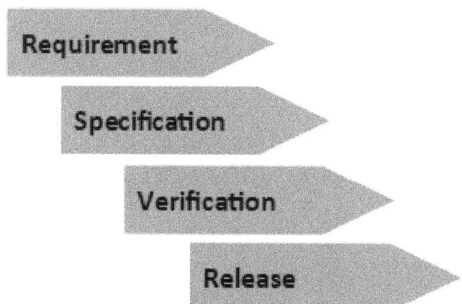

The above diagram introduces some key concepts that are applied during C&Q and validations. The intial stage deals with the requirements. Requirements should be specific, mesaable and unambigous. The requirements must decribe the use in my (intended use).

In response to requirements, a specification document is needed. The specifation document translates the user requirements into elements that help focus the contruction and design teams to meet the needs of the business. When contruction and assembly is completed the verificaiton activity can commence. Successful C&Q verification allows release of a system, factility or process for the subsequent validation activites and manufacturing.

1.2. Regulatory Requirements

If C&Q is to be applied within a pharmaceutical company, regulatory guidance in Europe is provided under Eudralex V4 Annex 15 Qualification and Validation.[1] The main stages include (i) user requirements (specification), (ii) Design Qualification (iii) Commissioning (iv) Installation Qualification, (v) Operational Qualification and (vi) Performance Qualification. The FDA and other regulatory bodies throughout the globe may require specific requirements to be considered for certain industries. For instance, 21 CFR 211.42(b) below states requirements covering the flow of materials through the building in a manner that minimizes contamination. Hence, the building design must consider the type of processes and manufacturing that will be undertaken. Design requirements, facility and utility requirements are required to be specified and determined to be appropriate for the intended use of the buildings.

> 21 CFR 211.42(b) states, that "The flow of components, drug product containers, closures, labelling, in-process materials, and drug products through the building or buildings shall be designed to prevent contamination."

21 CFR 211.42(c) states, in part, that "Operations shall be performed within specifically defined areas of adequate size. There shall be separate or defined areas or such other control systems for the firm's operations as are necessary to prevent contamination or mixups during the course of the following procedures:

---

[1] Eudralex Volume 4, EU Guidelines for Good Manufacturing Practice for Medicinal Products for Human and Veterinary Use Annex 15: Qualification and Validation

> Aseptic processing, which includes as appropriate:
>
> (i) Floors, walls, and ceilings of smooth, hard surfaces that are easily cleanable
>
> (ii) Temperature and humidity controls
>
> (iii) An air supply filtered through high-efficiency particulate air filters under positive pressure, regardless of whether flow is laminar or nonlaminar
>
> (iv) A system for monitoring environmental conditions
>
> (v) A system for cleaning and disinfecting the room and equipment to produce aseptic conditions
>
> (vi) A system for maintaining any equipment used to control the aseptic conditions."

As illustrated above, aseptic processing is achieved by design of facilities, control of environmental factors, monitoring and maintenance- and the integration of each one to establish and sustain parameters meeting aseptic requirements.

> 21 CFR 211.46(b) states that "Equipment for adequate control over air pressure, micro-organisms, dust, humidity, and temperature shall be provided when appropriate for the manufacture, processing, packing, or holding of a drug product."

Equipment such as HVACs and AHUs provide the air control parameters above within a cleanroom designed zone. It is noteworthy that the control of the environment and above parameters is applicable to manufacturing, processing, packaging and holding of drug products. Techniques that maintain sterility of sterile items and surfaces include:

• Ensuring that sterile materials only make contact with sterile instruments. Sterile instruments should always be used in the handling of sterilized materials. Between uses, sterile instruments should be held under Class 100 (ISO 5) conditions and maintained in a manner that prevents contamination (e.g., placed in sterilized containers). Instruments should be replaced as necessary throughout an operation. Re-sterilisation requirements of re-usable tools and instruments must be established and validated

After initial gowning, sterile gloves should be regularly sanitized or changed. This behavior helps to minimize the risk of contamination for the person transferring particles and micro-organisms from surface to surface. However, personnel should not directly contact sterile products, containers, closures, or critical surfaces with any part of their gown or gloves. Gloves should not make contact with personnels face or cleanroom clothing.

> 21 CFR 211.46(c) states, in part, that "Air filtration systems, including prefilters and particulate matter air filters, shall be used when appropriate on air supplies to production areas."

Personnel should also be educated to move slowly and deliberately, as rapid movements can create unacceptable air flow turbulence in a critical area and risks product sterility. If these movements occur they disrupt the unidirectional airflow, presenting a challenge beyond intended cleanroom design and control parameters. Personnel should keep their entire body out of the path of unidirectional airflow Unidirectional airflow design is used to protect sterile equipment surfaces, container closures, and product.

Personnel must maintain Proper Gown Control prior to and throughout aseptic operations. Only personnel who are qualified and appropriately gowned should be permitted access to the aseptic manufacturing area. The gown should provide a barrier between the body and exposed sterilized materials and prevent contamination from particles generated by, and microorganisms shed from, the body. All skin and hair should be covered. (face-masks, hoods, beard/moustache covers, protective goggles, and elastic gloves are examples of common elements of gowns). Written procedures should detail the methods used to don each gown component in an aseptic manner. An adequate barrier should be created by the overlapping of gown components (e.g., gloves overlapping sleeves). If an element of a gown is found to be torn or defective, it should be changed immediately. Gloves should be sanitized frequently.

> 21 CFR 211.63 states that "Equipment used in the manufacture, processing, packing, or holding of a drug product shall be of appropriate design, adequate size, and suitably located to facilitate operations for its intended use and for its cleaning and maintenance."

To maintain sterility of materials, a proper aseptic manipulation should be approached from the side and not above the product (in vertical unidirectional flow operations). Also, operators should refrain from speaking when in direct proximity to the critical area.

> 21 CFR 211.65(a) states that "Equipment shall be constructed so that surfaces that contact components, in-process materials, or drug products shall not be reactive, additive, or absorptive so as to alter the safety, identity, strength, quality, or purity of the drug product beyond the official or other established requirements."

Equipment used in aseptic processing must be made of suitable materials with appropriate means of assembly and surface finishes. High grade stainless steel is particularly suited as it suitable for cleaning with agents such as IPA and can withstand moist steam sterilization. There is no risk of leachables and does not shed material and does not generate particles. The mechanism to ensure appropriate materials and construction of equipment relies on establishing documentation on user requirements, design qualification and vendor contracts and agreements. This basis of design specifications can then be verified during factory acceptance testing, site acceptance testing and qualification.

> 21 CFR 211.67(a) states that "Equipment and utensils shall be cleaned, maintained, and sanitized at appropriate intervals to prevent malfunctions or contamination that would alter the safety, identity, strength, quality, or purity of the drug product beyond the official or other established requirements."

The prevention of equipment malfunction or contamination requires a schedule of preventative maintenance, regular cleaning and deep cleaning in order to maintain processes to safe and clean standards that support manufacturing quality products that are safe, pure, meet quality requirements and attributes such as identify and strength without contamination or unacceptable residuals.

> 21 CFR 211.113(b) states that "Appropriate written procedures, designed to prevent microbiological contamination of drug products purporting to be sterile, shall be established and followed. Such procedures shall include validation of any sterilization process."

It is necessary for manufacturers to regularly assess and audit conformance of personnel to relevant aseptic manufacturing requirements. The implementation of an aseptic gowning qualification program should assess the ability of a cleanroom operator to maintain the quality of the gown after performance of gowning procedures, with assessment to include microbiological surface sampling of several locations on a gown (e.g.glove fingers, facemask, forearm, chest). For any aseptic processing operation, if adverse conditions occur, additional or more frequent requalification could be indicated. To protect exposed sterilized product, personnel should to maintain gown quality and strictly adhere to appropriate aseptic techniques. Written procedures should adequately address circumstances under which personnel should be retrained, requalified, or reassigned to other areas.

1.2.1. C&Q Model

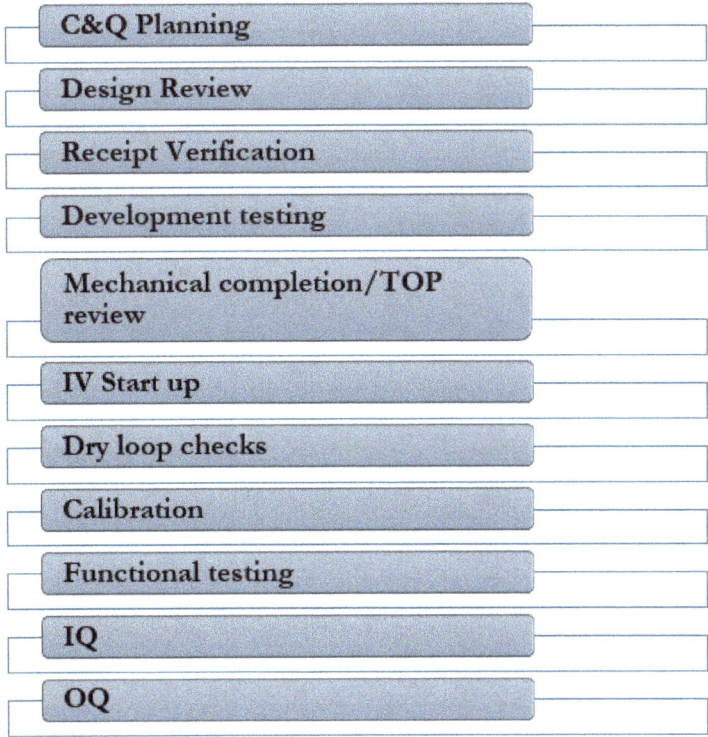

The above model is an example of a common approach to C&Q. The applicable model should be created by your company as the necessary product knowledge, quality, corporate and regulatory requirements are understood best by on which elements need apply.

1.2.2. Qualification Model for Manufacturing systems and Equipment

Further guidance on manufacturing systems and equipment in a pharmaceutical and biopharmaceutical context is issued by ASTM[2]

1.2.3. User requirements specification (URS)

A URS is a specification document that is written for equipment, facilities, utilities or systems that defines the specific requirements to meet the use application and intended use. This includes the physical, function, operational, electrical, performance requirements and so on. Creating a written URS ensures that the user requirements are documented and approved and this can be the basis for design qualification and providing a vendor or OEMs, Original equipment manufacturers with a 'build specification' The URS is an important input during the qualification and validation process and therefore should include all critical and quality related requirements.

1.2.4. Design qualification (DQ)

The next element in the qualification of equipment, facilities, utilities, or systems is Design Qualification, DQ where the compliance of the design with GMP should be demonstrated and documented. The requirements of the user requirements specification should be verified during the design qualification process. The DQ should be approved by relevant roles.

---

[2] ASTM, Standard Guide for Specification, Design and Verification of Pharmaceutical and Biopharmaceutical Manufacturing systems and Equipment

1.2.5. Commissioning

Commissioning includes the application of good engineering practices to introduce new equipment and facilities/utilizes into operation in a controlled manner that supports project delivery, safety and success.

- C&Q Planning
- Design Review
- Receipt Verification
- Development testing
- Mechanical completion/TOP review
- IV Start up
- Dry loop checks
- Calibration
- Functional testing
- IQ
- OQ

Commissioning is essentially a managed engineering approach to start up and provide turnover of equipment, utilities, systems and facilities. It involves field verification and review of system specific components and review of the construction, building and assembly of systems to ensure they meet the intended use and design specification. After the commissioning stage, these systems are then turned over to the responsible person or owner which can then proceed with qualification and validation as required. Therefore, the real value of an effective commissioning program is reducing risks in qualification and providing the basis of success.

C&Q planning is achieved by creating a C&Q plan which sets out the activites, deliverables and key information of a project. It should cover all of the elements listed and provide guidance to the C&Q team. It should maintained as accurate and updated if required periodically. Approval of the plan should include all stakeholders which provides a mechanism for agreeing the C&Q strategy and communicates the nuiances for each stage.

It is best practice to include a design review of systems/facilities that are been introduced. The design review involves reviewing the proposed design (drawings, specifications, materials, configurations) against pre-appproved requirements such as the URS and other procedures (company SOPs) and standards that may need to apply.

For a direct impact system a design review is normally completed under the term Design Qualification, DQ. Design Qualification should include a review of the following documents:
- List of the approved documents in scope of the review
- Scope of DQ/review
- Attendee list and function represented
- List of open items
- List of corrective actions
- Conclusions that Design is suitable for intended use and that the project may proceed to the next stage.

Receipt Verification

Receipt Verification (RV) ensures that components or systems and assocaited ancillary items and documentation is provided and received as purchased per vendor P.O.s. RV is performed prior to installation, assembly and verification activities. Therefore if an incorrect system or equipment is provided it can be returned to the provider with minimum delay. RV checks include:
- Confirmation of model number
- Inspection to include equipment is not damaged
- All specified items/components on P.O are fullfiled and reflected in delivery documentation.
  - Documentaition and manuals are available.

Receipt Verification Checklist

Project Code:		Location:	
System/Equipment		Model/Spec:	
Date of Verification:		Completed by:	

Receipt Verification (RV) ensures that components or systems and assocaited ancillary items and documentation is provided and received as purchased per vendor P.O.s. RV is performed prior to installation, assembly and verification activities. Therefore is an incorrect system or equipment is provided it can be returned to the provider with minimum delay. RV checks include:
- Confirmation of model number
- Inspection to include equipment is not damaged
- All specified items/components on P.O are fullfiled and reflected in delivery documentation.
    - Documentaition and manuals are available.

General Desc.	Action	Status	Reference	Result
P.O.	P.O. is available for review and meets requirements	Checked ☐ Not Checked ☐	PO20240786	Accept ☐ Reject ☐
Delivery	Confirm model number visually on equipment	Checked ☐ Not Checked ☐	Delivery note #237	Accept ☐ Reject ☐
Quantity	Order quantity is received	Checked ☐ Not Checked ☐	Delivery note #237	Accept ☐ Reject ☐
Damage	Visually inspect for visible damage	Checked ☐ Not Checked ☐	n/a	Accept ☐ Reject ☐
Ancillary items	Confirm ancillary items per P.O. are delivered and correct	Checked ☐ Not Checked ☐	Delivery note #237	Accept ☐ Reject ☐
Documents	Documentation is provided e.g. manuals, drawings, CE certification etc.	Checked ☐ Not Checked ☐	System manual X430	Accept ☐ Reject ☐
		Checked ☐ Not Checked ☐		Accept ☐ Reject ☐

Development testing

Development testing is the intial testing on a system that confirms that the system functions according to the design intent and the functions specifications. It can be completed in a simulated manner often termed 'off-line' testing. Common activities during this stage include I/O, (Input/ output)testing. Development testing, if documented accordingly following GEP and GDP standards and configuration controls may be leveraged to meet the requirements of functional testing or OQ, operational qualification testing

Mechanical completion/TOP review

Mechanical completion is confirmation that the system or equipment is phycically assembled and fixed in its use configuration. For stick built systems, drawings can be used to verify that the system is as designed and intended. Mechanical completion can also include construction activities. The contractor responsible for contruction and assembly signifies that the system is ready to handover to the C&Q team after Mechanical completion. The package of documentation is referred to as a turn-over package or TOP. Mechanical verification can occur prior to Installation verification (IV) as some checks and inspections may need access to restricted areas or areas subject to further modification. (e.g. pipr verifications prior to insulation). Open items or issues where the mechanical completion does not meet requiremetns or drawing specifications can be fixed in real time if feasible or alternatively may be tracked via a punch list. A punch list is a tracking mechanism where follow up actions, improvements or remediation is required prior to formal closure of the TOP package.

Piping - Mechanical Completion Checklist

Project Code:		Location:	
System/Equipment		Model/Spec:	
Date of Verification:		Completed by:	

A Mechanical completion checklist can be used as a screener for Mechanical Completion-Piping or as an additional control during execution of mechanical completion activity

1. Approved Drawings should be used during the completion of the Checklist.
2. Verify the below checks for all relevant locations
3. Complete a full walk through of the system

Desc.	Action	Status	Reference	Result
Components	Piping components installed per design specifications	Checked ☐ Not Checked ☐	Drawing xx Design spec	Accept ☐ Reject ☐
Piping	Piping and mechanical integration per drawings	Checked ☐ Not Checked ☐	Drawing xx	Accept ☐ Reject ☐
Connections	Inspect welded joints, flanges, threaded connections. Verify no leakage/gross defects	Checked ☐ Not Checked ☐	Drawing, Specifications	Accept ☐ Reject ☐
Valves and fittings	Confirm correct installation and operation and in correct position	Checked ☐ Not Checked ☐	Drawing Parts spec	Accept ☐ Reject ☐
Supporting structures	All supporting structures are per drawing and are secured	Checked ☐ Not Checked ☐	Drawing	Accept ☐ Reject ☐
Materials of construction	Materials conform to design specification	Checked ☐ Not Checked ☐	Drawing Design spec	Accept ☐ Reject ☐
Earthing and Crossbonding	Earthing and cross bonding per electrical drawing	Checked ☐ Not Checked ☐	Drawing	Accept ☐ Reject ☐
Identification	Pipework and valves are tagged and labelled	Checked ☐ Not Checked ☐	Drawing	Accept ☐ Reject ☐
Insulation	Insulation is appropriate and damage free	Checked ☐ Not Checked ☐	Design spec	Accept ☐ Reject ☐
Pressure testing	Pressure testing complete	Checked ☐ Not Checked ☐	Design spec	Accept ☐ Reject ☐
Documentation	All vendor and contractor documents available	Checked ☐ Not Checked ☐	Drawings, Certs, Manuals	Accept ☐ Reject ☐
Safety	Piping system has appropriate safety signs and controls	Checked ☐ Not Checked ☐	Certs and CE marking	Accept ☐ Reject ☐

- IV Start up
- Dry loop checks
- Calibration

Start up is the next step after installation verification is completed for each system. The safety of the system and its operation is important to verify, so start up is done in a controlled and safe manner. A start up protocol is best practice.

Installation Verification Checklist				
Project Code:		Location:		
System/Equipment		Model/Spec:		
Date of Verification:		Completed by:		
General Desc.	Action	Status	Reference	Result
Siting	Floor/siting is clean, level and appropriate for equipment/system	Checked ☐ Not Checked ☐	Manufacturers specification	Accept ☐ Reject ☐
Layout	There is adequate space bordering the equipment to allow safe operation	Checked ☐ Not Checked ☐	Layout drawing	Accept ☐ Reject ☐
Lighting	Adequate lighting is installed	Checked ☐ Not Checked ☐	URS	Accept ☐ Reject ☐
Packaging	All packaging, strapping, transport protection is removed	Checked ☐ Not Checked ☐	N/A	Accept ☐ Reject ☐
Mechanical Assembly	The equipment and ancillary is assembled (if required)	Checked ☐ Not Checked ☐	User manual	Accept ☐ Reject ☐
Anchoring	The equipment is bolted to the floor (if required)	Checked ☐ Not Checked ☐	User manual	Accept ☐ Reject ☐
Electrical connection	Verify electrical supply drop is situated in acceptable position	Checked ☐ Not Checked ☐	User manual	Accept ☐ Reject ☐
Pneumatic connection	Compressed air drop is available with regulator	Checked ☐ Not Checked ☐	User manual	Accept ☐ Reject ☐
Calibration	Initial calibration is completed according to standard	Checked ☐ Not Checked ☐	User manual	Accept ☐ Reject ☐
Safety	No sharp corners or hazards are identified	Checked ☐ Not Checked ☐	User manual	Accept ☐ Reject ☐
Documents	Vendor documentation available	Checked ☐ Not Checked ☐	User manual	Accept ☐ Reject ☐

Functional testing

The purpose of Functional testing (FT) is to verify the safety of components during operation. The activity can also be referred to as functional verification. It requires a combination of physical inspections and also completed tests in person by testing alarms, valves, interlocks, emergency stops and so on. FT of a cleanroom during the commissioning stage would include temperature and humidity monitoring, particulate monitoring and differential pressure monitoring.

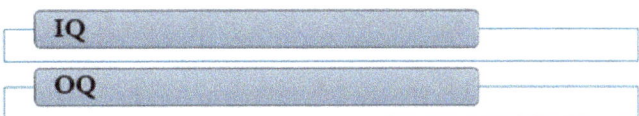

Installation qualification (IQ)

For the pharmaceutical, biotech and MedTech industries compliance to regulations of health authorities, competent authorities and notified bodies is required in order to manufacture and produce products. The regulators grant marketing authorizations which require companies to be subject to audit and inspection. Validation of processes, equipment, facilities and utilizes is mandated by various regulations, depending on the markets and geography. US Federal drug association require validation for both medical devices, pharmaceuticals, biotechnology, therapeutics and combination products. Installation Qualification, IQ should be performed on equipment, facilities, utilities, and systems, including computerized systems. In respect of manufacturing facilities, cleanrooms which are built-for-purpose environments designed to control particulate and climatic conditions. The type of equipment, facility or utility should inform the requirements of IQ. However, at a minimum the following should be considered:

- Verification of the correct installation e.g. pipe work and services as required by drawings and design specifications.
- Verification of the correct installation according to the supplier specifications and the intended use and purpose of the system.
- Calibration of instrumentation
- Verification of the materials of construction.
- Operating manual and supplier documentation
- maintenance requirements

Operational qualification (OQ)

OQ verification testing should include but is not limited to the following checks:
- Tests that have been developed from the knowledge of processes, systems and equipment to ensure the system is operating as designed
- Tests to confirm upper and lower operating limits, and /or "worst case" conditions

Completion of a successful OQ should allow the finalisation of standard operating and cleaning procedures, operator training and preventative maintenance requirements.

Performance qualification (PQ)

PQ should normally follow the successful completion of IQ and OQ.
PQ should include, but is not limited to the following:
- Tests, using production materials, qualified substitutes or simulated product proven to have equivalent behavior under normal operating conditions with worst case batch sizes.

The frequency of sampling used to confirm process control should be justified; ii. Tests should cover the operating range of the intended process, unless documented evidence from the development phases confirming the operational ranges is available for manufacturing.

Re-Qualification

Equipment, facilities, utilities and systems should be evaluated at an appropriate frequency to confirm that they remain in a state of control. Re-qualification is necessary and performed at a specific time period, the period should be justified and the criteria for evaluation defined. Furthermore, the possibility of small changes over time should be assessed.

2. EU GMP V4 Annex 15

2.1. Introduction

Annex 15 provides guidelines for using a risk-based approach for the qualification and validation of pharmaceutical processes including requirements for the validation of facilities, equipment, utilities, and computerized systems. Qualification of facilities should ensure they are suitable for their intended purpose, while the qualification of utilities should ensure they meet required specifications. The annex also discusses revalidation requirements and considers the circumstances necessitating revalidation. Other areas of guidance includes validation documentation, control and changes and requirements for clinical trials.

2.2. Changes to validated systems or processes

After a system or service is fully C&Q'd validation and in use for production purposes. Changes to the system should be evaluated and may require revalidation depending on the impact to the system. Replacement ancillary parts may not require validation and only maintenance intervention. However, maintenance intervention should be controlled, pre-approved and documented. A return to service checklist after any repairs of modifications is required. If upon evaluation, the change impacts the validated state of the equipment, then change management and change controls is required to preapprove the change and identify the actions needed to bring the system back into service while ensuring there is no adverse impact on the quality of the products or degradation of services provided. This detailed control and documentation is in compliance with Annex 15 which emphasizes the requirement for documentation and record-keeping throughout the validation lifecycle.

2.3. Validation and product lifecycle

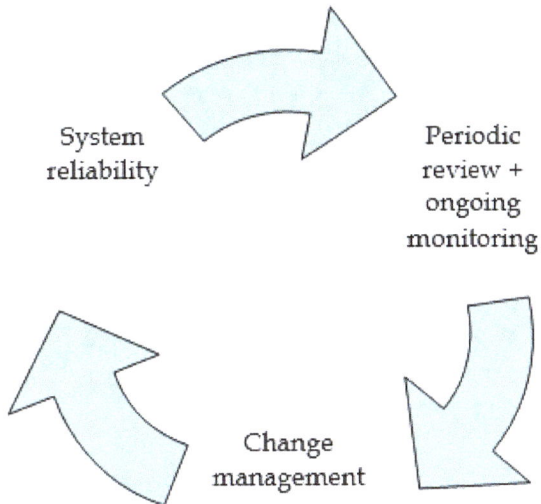

Annex 15 emphasizes the importance of ongoing monitoring and review of validated systems. Validated systems should be periodically reviewed to ensure continued compliance and effectiveness.

3. Cleanroom Qualification

3.1. Introduction

Cleanrooms are designed to maintain extremely low levels of viable particles, non-viable particles and gaseous fluid according to varying specifications, based on use. The levels of cleanliness and contamination control is based on the type of industry, specific requirements and industry and regulatory rules. Electronics and semiconductor manufacturing use cleanrooms as various stages of the process to ensure quality and functionality. These sensitive devices can be damaged or compromised by airborne contaminants and therefore are controlled in the manufacturing environment. Likewise, pharmaceutical manufacturing and biotechnology processes require very specialized levels of environmental control due to the impact temperatures, humidity and particulate can have on raw materials and products. A cleanroom is a specially constructed enclosed area has the following controlled parameters:

- Temperature
- Humidity (Relative Humidity)
- Sound and Vibration
- Lighting
- Airflow Pattern
- Pressurization
- Particle Count
- Microbial Contamination
- Gaseous Contamination

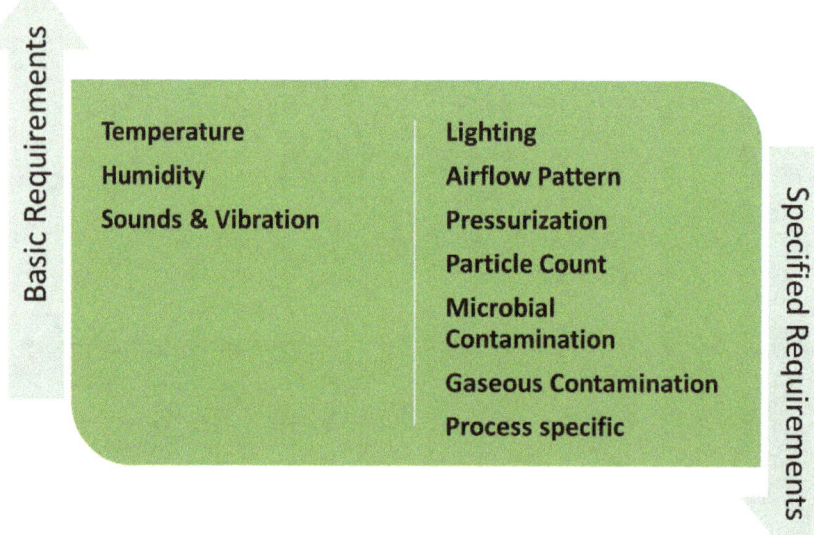

Heating, Ventilation and Air-Conditioning (HVAC) contributes to the functioning of clean zones. It works to prevent any negative effect on production due to changes in climatic conditions. In addition, it also works to prevent product contamination and providing ergonomic working conditions. Good Engineering Practices, application of standards, regulations and commissioning and qualification planning are necessary to deliver systems that are fit for purpose and perform as required.

- International Organization for Standardization (ISO), ISO 29463 - High-efficiency filters and filter media for removing particles in air, Parts 1 to 5.

- International Society for Pharmaceutical Engineering (ISPE) – Good Practice Guide – Heating, Ventilation and Air Conditioning (HVAC)
- International Organization for Standardization (ISO), ISO 14644 - Cleanrooms and associated controlled environments, Parts 1 to 9.
- FDA 21 CFR Parts 210 and 211 – Current Good Manufacturing Practice In Manufacturing, Processing, Packing or Holding of Drugs; General and Current Good Manufacturing Practice For Finished Pharmaceuticals
- PDA Technical Report No.13- "Fundamentals of a Microbiological Environmental Monitoring Program"
- EudraLex Volume 4, EU Guidelines for Good Manufacturing Practice for Medicinal Products for Human and Veterinary Use, Part 1, Chapter 3: Premises and Equipment)
- EN 1822 Series "High efficiency air filters (HEPA and ULPA)"
- EN 779 "Particulate air filters for general ventilation. Determination of the filtration performance."
- EN 1886 "Ventilation for buildings – Air Handling Units -Mechanical Performance"
- EN 12464-1 – "Light & Lighting of Indoor Work Places".
- ASHRAE Handbooks – Fundamentals, HVAC Systems and Equipment, HVAC Applications, Refrigeration
- ASHRAE Standard 110 – "Method of Testing Performance of Laboratory Fume Hoods"
- ASHRAE 52.2-1999 "Method of Testing General Ventilation Air-Cleaning Devices for Removal Efficiency by Particle Size"

3.2. Cleanroom Environment

The environment where products are manufactured, processed and packaging can lead to contamination issue that may impact the product and safety. Therefore, an appropriate environmental cleanliness level is required to minimize the risks of particulate or microbial contamination to the product. The levels of cleanliness depends on the activity and products been provided. A cleanroom is defined as enclosed area which is environmentally controlled with respect to particles, temperature, humidity, air pressure, air pressure flow patterns, air motion, vibration, noise, viable organisms, and lighting and is designed and constructed for the intended use in mind.

ISO 14644-1[3] defines a cleanroom as "a room in which the concentration of airborne particles is controlled, and which is designed, constructed and operated in a manner to control the introduction generation, and retention of particles inside a room".

ISO 14644-4 A.1[4] suggests clean rooms are "enclosed (rooms)or surrounded by further zones of lower cleanliness classification. This can allow the zones with the highest cleanliness demands to be reduced to the minimum size. Movement of material and personnel between adjacent clean zones gives rise to the risk of contamination transfer, therefore special attention should be paid to the detailed layout and management of material and personnel flow"

- Critical process areas are more stringently controlled portion of the cleanroom. Pharmaceutical cleanrooms and controlled zones should;
- Prevent the quality of products being impacted with unwanted airborne contaminants or particles and prevent products from contaminating each other
- Provide a comfortable environment for the operators and limit exposure to hazardous risks (e.g. drug particulates, fumes, vapors)
- Remove any contaminants form the room as effectively as possible and in accordance with regulatory requirements.

---

[3] ISO 14644-1 Cleanrooms and associated controlled environments Part 1: Classification of air cleanliness by particle concentration.
[4] ISO 14644-4 Cleanrooms and associated controlled environments — Part 4: Design, construction and start-up.

3.3. Cleanroom Zoning and Classification

Selecting a suitable classification for a room or manufacturing facility depends on several factors. Firstly, it can be said that sterile products require a more stringent set of criteria than non-sterile products. However, there is an extensive range of products and medical devices that are sterile but are used in different ways and consist of different materials and technology. Some sterile products are single use only and used for short-term purposes and then disposed of.

Other sterile products are used subcutaneously for longer periods or even require implantation. Therefore, the design of a facility along with its HVAC specification must be appropriate to the product being manufactured. High-risk products require greater control. The goal of facilities and HVAC systems is to minimise contamination and the associated risks. Using a sterile versus non-sterile rule of thumb is not adequate when classifying a room or facility.

Standards including EN ISO 14644-1 and guidelines such as EU cGMP Guidelines EudraLex volume 4 Annex 1 (2008) should be consulted in order to fully understand the requirements of each ISO classification and grade of room.

ISO classifications do not specify room occupancy states but when a designation is applied, the occupancy state must be stated in the relevant documentation or procedure. The most relevant European Guideline (Annex 1 of the EU cGMP Guideline) lists four classification grades and their associated particulate limits in the 'at rest' and 'in operation' conditions. In general, for the sterile and non-sterile products, similar classes are applied, but in non-sterile production the producer could assign their classes, having similar particulate concentration, temperature, pressure etc. but lower air-change rate could be used.

The classification of a cleanroom is determined by the maximum number of particles acceptable according to a specific size and per the volume of air. The selection of the right classification for any given cleanroom needs to consider the application, the type of products been processed and the type of processes. For example, a product that can be terminally sterilized generally requires less control and can be sealed in a area that is not fully aseptic. Particles in the air is made up of both Viable and Non-Viable Particles. Viable particles can present microbial risks. The levels of viable and non-viable particles is an indication of how 'clean' an area is. Therefore, monitoring these levels is useful in determining any adverse trends and tracking on an ongoing basis the cleanroom is operating as required.

Airborne particulates

- Particles larger than 100 microns can be seen with naked eye

- Particles ranging from 0.001 to 100 microns were the main particles of interest in relating to contamination

- Atoms and molecules used to be considered too small but not anymore due to the recognition of molecular contamination

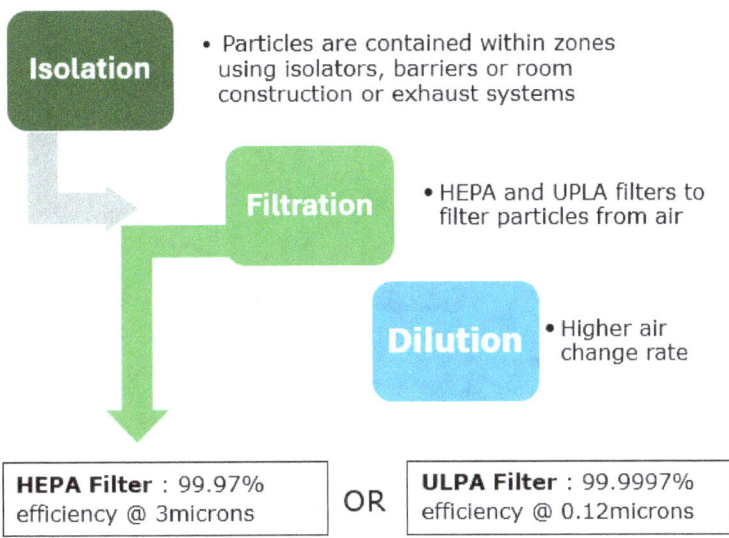

Critical areas such as ISO Level 5 are afforded protection by areas of a lower classification. Raw materials, components and personnel are controlled with an increasing level of cleanliness in order to prevent contamination from the outside impacting the critical zones.

3.4. Types of Contamination

- cross contamination (of a product/material with another product/material)
- non-microbial particulate contamination (non-viable particles)
- biological/microbiological contamination (viable particles/micro-organisms)

Factors Influencing Contamination Cleanliness Levels in the Manufacturing Processes:
- process
- air cleanliness
- personnel hygiene and clothing
- work practices
- material design (material of construction, surface finishes, room finishes, equipment, open system/enclosed system, utensils etc.)
- material cleanliness

3.5. Cleanroom Classification Table

The maximum particle levels in per ISO 14644-1, both "At Rest" and "In Operation" particle levels are indicated below:

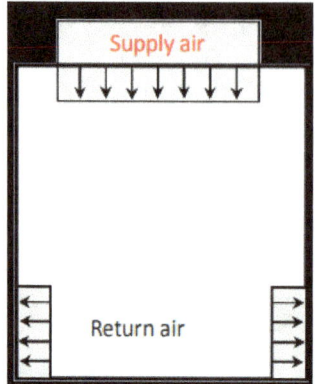

As built is the condition where the installation is complete with all services connected and functioning but with no production equipment, materials or personnel present.

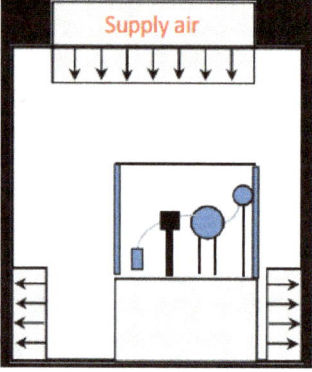

At rest condition is where the installation is complete with equipment installed and operating in a manner agreed upon by the customer and supplier, however, no personnel present.

In operation is the condition where the installation is complete with equipment installed and operating in a manner agreed upon the customer and supplier and where personnel present and working.

3.6. Zone Classification

Applying ISO 14644- 1 rules, cleanrooms are classified based on the level of airborne particulates within the environment. ISO Class 5-9 are summarized below, with ISO Class 9 allowing the greatest levels of particulate.

ISO ZONE 5

Critical zones are areas or clean room zones where the product, packaging, or closures are exposed to environmental conditions during the completion of the last manufacture steps. This exposure to the environment may impact product quality. This control of the critical zone is achieved by the design of the room, use of HEPA filtration during HVAV, gowning requirements, access control and the control and monitoring of conditions such as temperature, relative humidity and pressure differentials.

ISO Class 5 permits a maximum allowable particles per cubic meter: 3,520
Airborne particulates are limited to a very low level, making it suitable for environments requiring extremely high levels of cleanliness, for example, pharmaceutical production and Aseptic manufacturing and biotech.

ISO Zone 6

ISO Class 6 permits a maximum allowable particles per cubic meter of 35,200. Therefore, this cleanroom has a higher particle count compared to ISO 5 cleanrooms but are still maintained to high cleanliness standards.

ISO Zone 7

The maximum allowable particles per cubic meter is 352,000
ISO Class 7 cleanrooms have higher particle counts compared to Class 6

ISO Zone 8

ISO Class 8 permits a maximum allowable particles per cubic meter of 3,520,000. ISO Class 8 cleanrooms have even higher particle counts compared to Class 7 and are considered as controlled environments with a lower degree of cleanliness E.G. food processing, or certain medical device manufacturing processes.

ISO Zone 9

ISO Class 9 permits the maximum allowable particles per cubic meter of 35,200,000. ISO Class 9 cleanrooms are used when high levels of cleanliness is not critical. ISO classifications ensure that cleanrooms function at the appropriate levels of cleanliness to support the operations and activities completed within them.

3.7. HVAC Particulate Control

The main purpose of the HVAC system in a cleanroom is to ensure the processing environment does not negatively impact upon product quality. Prior to the design and specification of a HVAC system, the product(s) and processes need to be understood to assess and determine the environmental controls necessary for a particular product, taking into account the type of product, the product specification and packaging and the regulatory requirements of competent authorities and notified bodies.

3.8. Total Airflow Volumes & Recovery Rates

Air change rates per hour, (AC/hr) are an important factor in contamination control. The arbitrary 20 AC/hr are a result of previous industry standards, however, nowadays, the number of air changes and depends on several factors including:
- Particle Generation Rate, (PGR) inside the space from people, equipment, etc.
- Room supply air volume
- Quality of air distribution (ventilation efficiency)
- Cleanliness of dilution air (negligible if HEPA filter are used)

3.8.1. Particle Generation Rate (PGR)

PGR is a measure of the number of viable and non-viable particles being generated in the cleanroom from both people and equipment and to a lesser extent the building fabric as it should be designed to be non-shedding. Good cleanroom gowning and personnel training are an important factor in reducing room particle levels and AC/hr.

3.8.2. Room Supply Air Volume

The supply air volumetric flowrate to a room is not only determined by a required particle level in the room but also by several other interrelated factors:

- Room heat gains (internal and external)
- Number of occupants in the space and activities
- Gowning levels
- Moisture gain to the space from internal and external influences
- Room leakage and differential pressure requirements

Heat and humidity gain are typically more easily controlled but should be considered by the HVAC designer. Particle generation and removal is generally the main driver of the supply air volume and hence air change rates in cleanrooms. HVAC designers default to "rules of thumb" for supply air rates by class of space, rather than calculating the actual airflow rate based on the activities in the room.

3.8.3. Non-unidirectional flow & unidirectional

Non-unidirectional flow uses air turbulence and dilution to mix particle contamination generated by people and machinery in the clean room. Clean Filtered air is delivered to the room through ceiling mounted air diffusers. This air mixes with the room air and removes airborne contamination through air extracts generally at low level in the walls.

For large rooms swirl diffusers induce room air vertically up to the diffuser to mix with the supply air. These diffusers create good dilution of contaminants in the room over the perforated diffuser type and may be used in rooms where there is minimum dust liberation. However, they should be avoided in rooms where excessive dust is generated as they would add to the distribution of the dust and could be hazardous for the operators.

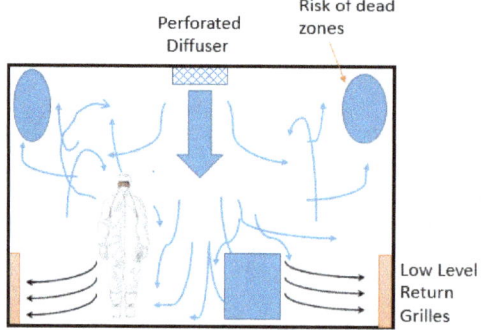

Non-Unidirectional flow with Perforated Diffuser

A shortcoming of non-unidirectional cleanrooms is the creation of air dead zones with high particle counts.

These pockets can persist for a period of time, and then disappear. This is due to currents that are set up within the room due to process related activity combined with the random nature of the downward airflow. Airflows should be planned in conjunction with operator locations, to minimize contamination of the product by the operator.

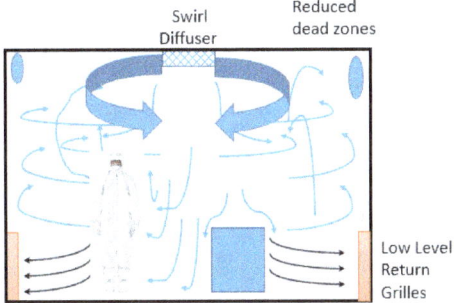

3.9. Unidirectional flow

Unidirectional airflow (aka not laminar) is defined by ISO14644-1 as a "controlled airflow through the entire cross-section of a cleanroom or a clean zone with a constant average velocity steady velocity and air streams that are considered to be parallel".

Unidirectional airflow is achieved by supplying filtered air through 95-100% of the ceiling. The air moves vertically downward laterally from the ceiling to return air grilles at low level.

This approach allows the contamination generated by the process or surroundings to drift to the floor level where they are extracted. This is known as the displacement method as it develops minimum air turbulence.

Unidirectional airflow velocity should be uniform and sufficient to dilute and remove particles generated in the room before they settle on a surface.

The particles are finally captured by the low-level return grilles and returned through the return walls and recirculated through the filters in the AHUs or ceiling Fan Filter Units (FFU). Cleanrooms with classification rating Zone 5 or below are almost invariably designed for unidirectional airflow.

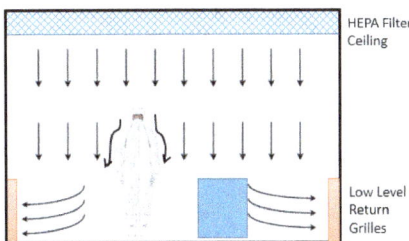

Unidirectional (turbulent or dilution) Airflow

3.9.1. Airlocks

Airlocks are small rooms (typically anterooms or material transfer) with interlocked doors, constructed to maintain airflow gradient and air pressure control between adjoining rooms (generally with different air cleanliness standards).

The primary role of an airlock is to provide a pressure buffer between areas of different classification, and for that reason its own internal pressure is somewhere between (floats), or equal to, the rooms it connects.

If a door is opened, the adjoining rooms will be at the same pressure, and contamination can flow across the doorway, most easily moved by personnel or equipment dragging the contamination along.

Therefore, once the doors are closed, there must be enough air changes to reduce the contamination level before the next door opens (this is one of the few places where air changes are important (quick room recovery).

So, the higher the air change rate in the airlock, the faster the recovery and the less chance of contamination passing into the next room.

Recovery to zero counts really can't happen with personnel in the room, but the airlock presents a relatively SMALL volume of lightly contaminated air that may be dragged through the next door to be opened.

Airlocks are generally divided into two categories: Personal Airlocks (PAL) and Materials Airlocks (MAL), these control personal and material flow into and out of clean spaces through a series of interlocking doors. Airlocks also help to maintain room pressure differentials between rooms of different classifications.

There are typically three (3 No.) types of airlock pressure arrangements used:

1. Cascade - airflow from areas of higher pressure, through the airlock to the area of lower pressure.
2. Bubble – airlock is at highest pressure to surrounding rooms, air flows from the airlock to the clean rooms.
3. Sink – airlock is at lowest pressures to surrounding rooms, air flows from the clean room and corridors.

Sink or Bubble configuration are only used to restrict cross contamination of products between rooms. Cascade is the typical arrangement used where the airlock pressure floats between the pressure in both rooms.

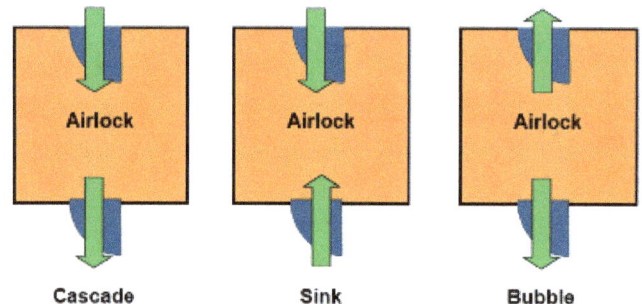

Airlock pressure configurations

Door interlocks: An interlock system or visual or audible alarm system is recommended to prevent the opening of more than one of the airlock doors at a time. Interlocked doors prevent an air flow through the airlock. Interlocks should be disarmed in case of emergency.

Pass-throughs: Small material airlock, called pass-throughs (PT) see Photo 7.2, which are too small for personnel, are used to transfer product from the higher-class rooms to lower class areas. Pass-throughs usually have interlocking sliding doors to also maintain clean room pressures between two different zones.

PT fall into two categories, namely dynamic or passive. Passive PT is typically used in cleanrooms with interlocked doors. PT shall be big enough to receive small items for example batch cards, samples, small consumables. Room Differential Pressure **(ΔP)**

As most facilities consist of multiple rooms with different requirements for cleanliness, differential pressure is required between the cleanrooms to ensure airflow from the cleanest spaces to the least clean spaces.

Certain operational activities may require a pressure differential to be maintained between rooms with the same classification but require an air pressure cascade (e.g. Autoclave Loading vs. Unloading). Where required this pressure must be less than the difference between room of different classification, typically +5 Pa. The requirement is to be determined at design stage and verified at certification.

The pressure differential of a cascade airlock is measured across the airlock and not across each door. Therefore, when only one door of an airlock is opened, a measurable DP between the cleanrooms persists. It also ensures the room pressure of the highest cleanliness room is maintained at a reasonable level.

Room Construction: Hard-ceiling construction is preferred for pressure-controlled spaces. In addition, air migration above the ceiling should be minimized between controlled and uncontrolled spaces.

To maintain pressurization within rooms, all doors should be fitted with continuous seals, manufactured of materials acceptable for cleanroom operation and wipe down. The gap between the finished floor and the bottom of door should be uniform at approximately 5mm (3/16 inch) when closed. Door floor sweeps are not recommended for swing doors due to their accumulation of dirt, scratching of floor, and maintenance. Doors preferably should operate, such that the pressure differential pushes the doors closed against the frame. Should the doors open to the low-pressure side, the door closer springs should be sufficient to hold the door closed and prevent the pressure differential from pushing the door open resulting in excessive leakage.

Air Leakage: Care shall be incorporated during design and construction to eliminate air ex-filtration within classified and controlled spaces to reduce air make-up and to maintain better static pressure control. The following, which is non-exhaustive, identifies areas where leak occurrences are most prevalent:

- Ductwork & Pipework penetrations through walls and termination into rooms
- Door perimeter
- Door closure mechanisms
- Surface Interfaces
- Spaces around equipment
- Access panels
- Electrical Fittings
- Light fixtures

The airflow leakage rate should be calculated for each room. This calculation should be based on the known architecture and the design pressure differential established for the facility

Differential Pressure Monitoring: For classified cleanrooms, it is recommended that pressure differential between cleanrooms be monitored and recorded continuously throughout each shift. There are 2 methods of measurement commonly used to monitor room pressure:

- Room to Room – Differential room Pressure
- Room to common reference point

Room to Room measurement directly meets the GMP requirement for room DP monitoring which clearly indicates the pressure relative to an adjoining room. This method is preferred for monitoring only i.e. by an EMS. When automatic room pressure control is used (VAV), the preferred measurement is room to a common reference point for stability. There is no GMP requirement that room pressure or airflow is automatically controlled but it is recommended in clean room areas to ensure pressures are maintained. A BMS is used for automatic control and monitoring and non-critical alarming. Fixed/manual damper control (CAV) on the supply air or return air is not recommended for cleanrooms due to its inability to compensate for fluctuations in supply and return fan output, filter loading, and exhaust modulations.

Airflow variations from dust collecting, vacuums, or process systems, and their effect on space pressurization, should be accounted for in the operation of the HVAC system.

3.9.2. Personnel considerations for sterile operations

Human interactions within an environment gives rise to contamination and the generation and movement of particles, this is due to the persons themselves (skin, hair etc.), particles that they carry on them and their clothing, and the clothing itself As a starting point of sterile operations, the design of the area and processes should minimize personnel intervention. A general rule of thumb applies where operator activities increase in a sterile or aseptic processing operation, the risk to finished product sterility also increases. The use of aseptic technique at all times mitigates the risk to patient and product. Aseptic technique must be followed at all times during the processing of aseptic products until they are packaged and closed appropriately. Appropriate training should be conducted before an operators or personnel are permitted to enter an aseptic manufacturing area. The basis of training should educate personnel on the requirements and reasoning. Practical instruction where applicable should also be provided. Areas such as aseptic technique, cleanroom behavior, microbiology, hygiene, gowning, patient safety and putting the patient first. The importance of written procedures should be highlighted with training to for each person relevant to their job and responsibilities.

> **21 CFR 211.22(a) states that "There shall be a quality control unit that shall have the responsibility and authority to approve or reject all components, drug product containers, closures, in-process materials, packaging material, labelling, and drug products, and the authority to review production records to assure that no errors have occurred or, if errors have occurred, that they have been fully investigated. The quality control unit shall be responsible for approving or rejecting drug products manufactured, processed, packed, or held under contract by another company."**
>
> **21 CFR 211.22(c) states that "The quality control unit shall have the responsibility for approving or rejecting all procedures or specifications impacting on the identity, strength, quality, and purity of the drug product."**

> 21 CFR 211.25(a) states that "Each person engaged in the manufacture, processing, packing, or holding of a drug product shall have education, training, and experience, or any combination thereof, to enable that person to perform the assigned functions. Training shall be in the particular operations that the employee performs and in current good manufacturing practice (including the current good manufacturing practice regulations in this chapter and written procedures required by these regulations) as they relate to the employee's functions. Training in current good manufacturing practice shall be conducted by qualified individuals on a continuing basis and with sufficient frequency to assure that employees remain familiar with CGMP requirements applicable to them."

Achieving and maintaining an Aseptic zone can be realized by a combination of facility design, material flows, engineering controls (Positive pressure, HVAC etc.), storage of materials, work practices, aseptic technique and proper planning of the tools and equipment required at the appropriate times and at the appropriate locations. Some of these techniques are described below.

Aseptic Technique: Contact sterile materials only with sterile instruments Sterile instruments should always be used in the handling of sterilized materials.

- Between uses, sterile instruments should be held under Class 100 (ISO 5) conditions and maintained in a manner that prevents contamination (e.g., placed in sterilized containers).

- Instruments should be replaced as necessary throughout an operation. After initial gowning, sterile gloves should be regularly sanitized or changed, as appropriate, to minimize the risk of contamination. Personnel should not directly contact sterile products, containers, closures, or critical surfaces with any part of their gown or gloves.

- Move slowly and deliberately Rapid movements can create unacceptable turbulence in a critical area. Such movements disrupt the unidirectional airflow, presenting a challenge beyond intended cleanroom design and control parameters. The principle of slow, careful movement should be followed throughout the cleanroom.

- Keep the entire body out of the path of unidirectional airflow Unidirectional airflow design is used to protect sterile equipment surfaces, container closures, and product. Disruption of the path of unidirectional flow air in the critical area can pose a risk to product sterility.

- Approach a necessary manipulation in a manner that does not compromise sterility of the product

- To maintain sterility of nearby sterile materials, a proper aseptic manipulation should be approached from the side and not above the product (in vertical unidirectional flow operations). Also, operators should refrain from speaking when in direct proximity to the critical area.

- Maintain Proper Gown Control Prior to and throughout aseptic operations, an operator should not engage in any activity that poses an unreasonable contamination risk to the gown. Only personnel who are qualified and appropriately gowned should be permitted access to the aseptic manufacturing area.

- The gown should provide a barrier between the body and exposed sterilized materials and prevent contamination from particles generated by, and microorganisms shed from, the body.

- The Agency recommends gowns that are sterilized and non-shedding, and cover the skin and hair (face-masks, hoods, beard/moustache covers, protective goggles, and elastic gloves are examples of common elements of gowns).

- Written procedures should detail the methods used to don each gown component in an aseptic manner. An adequate barrier should be created by the overlapping of gown components (e.g., gloves overlapping sleeves).

- If an element of a gown is found to be torn or defective, it should be changed immediately. Gloves should be sanitized frequently.

- For any aseptic processing operation, if adverse conditions occur, additional or more frequent requalification could be indicated. To protect exposed sterilized product, personnel should to maintain gown quality and strictly adhere to appropriate aseptic techniques. Written procedures should adequately address circumstances under which personnel should be retrained, requalified, or reassigned to other areas.

> 21 CFR 211.25(b) states that "Each person responsible for supervising the manufacture, processing, packing, or holding of a drug product shall have the education, training, and experience, or any combination thereof, to perform assigned functions in such a manner as to provide assurance that the drug product has the safety, identity, strength, quality, and purity that it purports or is represented to possess."
>
> 21 CFR 211.25(c) states that "There shall be an adequate number of qualified personnel to perform and supervise the manufacture, processing, packing, or holding of each drug product."
>
> 21 CFR 211.28(a) states that "Personnel engaged in the manufacture, processing, packing, or holding of a drug product shall wear clean clothing appropriate for the duties they perform. Protective apparel, such as head, face, hand, and arm coverings, shall be worn as necessary to protect drug products from contamination."

- Supervisory personnel should routinely evaluate each operator's conformance to written procedures during actual operations. Similarly, the quality control unit should provide regular oversight of adherence to established, written procedures and aseptic technique during manufacturing operations.

- There should be an established program to regularly assess or audit conformance of personnel to relevant aseptic manufacturing requirements. An aseptic gowning qualification program should assess the ability of a cleanroom operator to maintain the quality of the gown after performance of gowning procedures.

- An assessment should examine microbiological surface sampling of several locations on a gown (e.g., glove fingers, facemask, forearm, chest). Sampling sites should be justified.
- Following an initial assessment of gowning, periodic requalification will provide for the monitoring of various gowning locations over a suitable period to ensure consistent acceptability of aseptic gowning techniques. Annual requalification is normally sufficient for those automated operations where personnel involvement is minimized and monitoring data indicate environmental control.

> 21 CFR 211.28(b) states that "Personnel shall practice good sanitation and health habits."
>
> 21 CFR 211.28(c) states that "Only personnel authorized by supervisory personnel shall enter those areas of the buildings and facilities designated as limited-access areas."
>
> 21 CFR 211.28(d) states that "Any person shown at any time (either by medical examination or supervisory observation) to have an apparent illness or open lesions that may adversely affect the safety or quality of drug products shall be excluded from direct contact with components, drug product containers, closures, in-process materials, and drug products until the condition is corrected or determined by competent medical personnel not to jeopardize the safety or quality of drug products. All personnel shall be instructed to report to supervisory personnel any health conditions that may have an adverse effect on drug products."

- The basic principles of training, aseptic technique, and personnel qualification in aseptic manufacturing also are applicable to those performing aseptic sampling and microbiological laboratory analyses. Processes and systems cannot be considered to be in control and reproducible if the validity of data produced by the laboratory is in question.
- The quality control unit should establish a more comprehensive monitoring program for operators involved in operations which are especially labor intensive (i.e., those requiring repeated or complex aseptic manipulations).
- Asepsis is fundamental to an aseptic processing operation. An ongoing goal for manufacturing personnel in the aseptic processing room is to maintain contamination-free gloves and gowns throughout operations.
- Sanitizing gloves just prior to sampling is inappropriate because it can prevent recovery of microorganisms that were present during an aseptic manipulation. When operators exceed established levels or show an adverse trend, an investigation should be conducted promptly.

- Follow-up actions can include increased sampling, increased observation, retraining, gowning requalification, and in certain instances, reassignment of the individual to operations outside of the aseptic manufacturing area. Microbiological trending systems, and assessment of the impact of atypical trends, are discussed in more detail under Section X. Laboratory Controls.

> 21 CFR 211.42(c) states, in part, that "Operations shall be performed within specifically defined areas of adequate size. There shall be separate or defined areas or such other control systems for the firm's operations as are necessary to prevent contamination or mix-ups during the course of the following procedures: Aseptic processing, which includes as appropriate:
>
> "A system for monitoring environmental conditions ."
>
> 21 CFR 211.113(b) states that "Appropriate written procedures, designed to prevent microbiological contamination of drug products purporting to be sterile, shall be established and followed. Such procedures shall include validation of any sterilization process."

3.10. Room Temperature and Relative Humidity

The ratio of the actual water vapour pressure of the air to the saturated water vapour pressure of the air at the same temperature expressed as a percentage. More simply put, it is the ratio of the mass of moisture in the air, relative to the mass at 100% moisture saturation, at a given temperature. The normal operating temperature requirement for each classification .Temperature and humidity must be appropriate to the product and process. Consideration should be made for specific product and process requirements.

A facility should meet stated relative humidity design conditions; however, the acceptability of the facility or operation depends on meeting the operating ranges.

Room Temperature and Relative Humidity (RH) requirements depend on the product requirements and operator comfort. A risk assessment should be completed to determine the criticality of temperature and humidity on product quality.

Room temperature and RH and determined both are not critical parameters as the majority of product are elastomers whose impact temperatures have a wide band. Therefore, Temperature and Humidity should only be monitored and controlled for human comfort and kept within a range to ensure human discomfort (e.g. perspiring or dehydration) does not indirectly impact product quality.

As Temperature and RH are considered non-critical parameters they should be monitored and controlled on the non-validated Building Management System (BMS). If a Risk Assessment determines that Temperature or RH are critical parameters that may impact product quality, then these parameters should be monitored by a validated Environmental Monitoring System (EMS).

It is important to define the distinction between the design and operating parameters of a space. In Figure 7.10 below, Values of Critical Parameters of a Product indicates the relationship between the design, normal operating, and qualified (validated) operating (product stability range) ranges.

3.11. Temperature

Room temperature is maintained for personnel comfort, considering working activities, and should take into consideration the various gowning levels worn in each area to ensure the majority of personnel are comfortable. Energy usage should also be considered when selecting room temperature verses levels of gowning. Most HVAC systems have a lower limit of 16°C & 75%RH without additional expensive HVAC equipment with higher running costs. The temperature sensor should be specified with an accuracy ± 1°C (± 2°F) and be fully adjustable and calibrated annually.

3.12. Relative Humidity (RH)

Room Relative Humidity (RH) for personnel comfort should consider working activities to evaluate and determine the normal operating range and alert limits, while the product environmental requirements, if any, determine the qualified operating range where RH may have an impact on product. Where relative humidity levels are not specified as having product impact (e.g. aqueous product) the outer limits of the operating range or validation acceptance criteria shall be based on the following; In areas with a requirement for low particles and dust generation (e.g. clean room) the humidity should be maintained above 30% to minimize the risk of static electricity and particle generation due to dryness. Also, liquid products can lose moisture to a low-humidity space/room over an extended period. The risk of condensation and microbial growth increase above 70%RH. As a result, the Lower (Low-Low) and Upper (High-High) outer limits of the operating range or validation acceptance criteria (Alarm Limits) should be set at 30% and 70% in classified areas. For Warehouse areas humidity levels should be maintained between 0-90%RH.

3.13. Control and Spread of Smoke

Systems shall not encourage the spread of smoke and fire, and in some instances, may be required to provide positive control. Careful attention must be paid to how smoke will be controlled and eliminated. It is important that smoke levels be quantified, with the necessary containment level established (i.e., cfm [m/s] smoke passage through the required smoke barrier), based on the type of structure in question, the characterization of the occupants, and their expected time to egress. Examples of positive control include pressurization of escape routes and smoke venting systems.

3.14. Cleaning

The selection of cleaning methods for cleanrooms and the sited equipment should be confirmed early in the design process because the selection may affect other design features (e.g. construction and finishes, cleanroom layouts, auxiliary services, etc.). Effectiveness of cleaning should be addressed in the validation, as applicable. Physical cleaning should be controlled by procedure and be recorded as specified.

4. HVAC Systems

4.1. Introduction

Heating, ventilation and air-conditioning (HVAC) provide a critical function in the manufacturing of medical devices, pharmaceutical and biotech products by contributing to the quality and environmental conditions during manufacturing, processing and packaging. Temperature, relative humidity and ventilation should be appropriate and should not adversely affect the quality of pharmaceutical products during their manufacture and storage, or the accurate functioning of equipment.

Design parameters and user requirements should, therefore, be set realistically for each project, with a view to creating a cost-effective design, yet still complying with all regulatory standards and ensuring that product quality and safety are not compromised.

4.2. HVAC System Design

The HVAC system must be appropriately selected using the specific design requirements as outlined above. The system must be able to provide clean, conditioned air to the specified areas to meet all of the quality requirements. The most important precursor to HVAC design is the comprehensive definition of the function and performance required followed by the selection of an appropriate system. A poor selection can lead to unnecessarily high-energy consumption, and operational deficiencies. All-air systems rely on the movement of large quantities of air through a central air handling unit to control room conditions, as well as provide for ventilation requirements.

They have the advantage of being relatively simple with most of the unit situated in one location; however, they are very space consuming. All-air systems tend to be relatively inflexible and not ideal for areas that are likely to need environmental alteration on a regular basis.

These HVAC systems are used for areas that have a lot of small zones, each with slightly different thermal loads but which requires constant ventilation. These systems can have poor energy efficiency if a lot of reheat is required. These are typically used in large manufacturing areas, and laboratories with many small rooms.

HVAC systems are typically situated above production areas, though this isn't always the case. Air Handling Units (AHUs) are usually located on technical floors. Air is distributed through various channels:

- Above false ceilings
- Through shafts
- Through double-wall clean room walls

Cleanrooms may have more than one classification shared among adjoining suites, depending upon manufacturing, research and development, and containment requirements. The cleanroom, or main assembly area, shall be specifically designated either as a specific ISO classification or Controlled Not Classified (CNC) classification, however, adjoining spaces may be designated an alternate class, and controlled via differential pressure requirements.

Other factors can affect the environmental conditions within the CR and/or CNC. Examples of these factors include the following: Number of personnel occupying each area Number and types of equipment Cleaning frequency (e.g. equipment and facility) Personnel gowning Airflow (e.g. directional, turbulent, and rate per hour) Training (e.g. movement, behavior, hygiene) Factors, such as these, may affect the cleanroom system and should be considered in the design criteria and prescribed in the user requirement specification document(s).

A well designed environment is constructed with materials that allow for ease of cleaning and sanitization. Current Good Manufacturing Practices (cGMPs) require that buildings be of suitable size, construction, and location to facilitate cleaning, maintenance, and proper operation. Additionally, the cGMPs are concerned with the potential contamination and cross contamination of product. Based on the environmental needs of the product and/or process, controlled environment cleanrooms and areas are designed to separate manufacturing operations, and minimize the potential for contamination. The category and level of contamination control required by the product will help determine the Abbott Vascular room categorization. Abbott Vascular room classifications are based on non-viable air particulate requirements during At Rest conditions. Table 1 reflects minimum environmental specifications by Room Classification.

HEPA filters and Dehumidification

For most HVAC applications, dehumidification is best achieved by the use of cooling coils. It should be noted that dehumidification is a very high consumer of energy and should only be used if there is a real process need. When areas are not in use, the dehumidifier should be turned off, if possible.

When room humidity must be maintained below 50% during warm weather, an absorption dryer may be necessary unless the room temperature can be increased within specification to compensate. Normal practice is to optimise size and efficiency of the absorption dryer by first removing as much moisture from the air as possible by cooling. The design of absorption dryers is normally based on a slowly rotating desiccant wheel.

Air is passed through the wheel and dried by the desiccant coating (guidance: lithium chloride especially if the wheel is not used frequently and silica gel if used permanently and with low humidity target). It is not normally necessary to size a dryer to handle the entire air volume. Drying a proportion of air and re-mixing to achieve the desired moisture content is usually sufficient.

Air humidification may be necessary during cold weather when introducing fresh air to spaces that require humidity control. When air humidification is necessary, humidifiers should be selected on the following basis:

- direct steam injection using steam
- direct steam injection using self-generative electric or gas steam humidifier

Humidifiers should be located before the fan and the final filter which will remove any particulate generated. At least 300 mm clearance should be allowed upstream and 1 m downstream between humidifier manifolds and coils, attenuators etc. (general recommendation to be confirmed through calculation note provided by the vendor). A single manifold or multiple manifolds in parallel may be used to meet the humidification requirements as per manufacturer's recommendations.

Sound Attenuators

Sound attenuators should be provided as necessary, to achieve the specified noise levels within occupied spaces. To minimise external noise nuisance, assessment can confirm the necessity to use acoustic media (enveloped in polyester film), that is inert and corrosion-resistant at normal operating conditions. Material quality shall be equivalent to that specified for HVAC unit or ducts. Sound attenuators should be installed in the air handling unit or ductwork. The use of sound attenuators in the air supply and air return should be based on requirements for fresh air inlet and air exhaust, and according to external noise levels that might need to be maintained at or below the ambient site noise levels.

Dampers

The provision of sufficient dampers is essential for proper control. To minimise noise transmission into the room, these should be mounted as far as possible from the diffuser. Carefully evaluate the space-by-space pressure control that will be used in the design. Static pressure control via hard balance or dynamic control via air terminal control units are both appropriate. Consideration should be given to the overall project size, the complexity of the facility and the project budget.

Automatic volume controllers are recommended for regulating air volume independently of supply pressure. They can be selected for constant volume, variable volume or dual duct mixing applications. Automatic low-leakage fresh air and exhaust air shutoff dampers are strongly recommended to isolate the HVAC network. Fresh air dampers shall be Class 3 minimum (maximum leakage preventing coil freezing). Whenever fumigation is performed shutoff damper shall ensure Class 4 leakage rate. Where dampers are required to provide modulating control of airflow, they must be selected to provide an appropriate level of control authority. This will normally mean a damper smaller than the duct size.

Heating and Cooling

Heating mode: Low pressure hot water (LPHW) is the preferred heating medium for HVAC applications and should be used whenever practicable. Electrical heating should be avoided due to fire risk and should be limited to low power coil and in locations where no other energies are available. Hazard operability analysis (HAZOP) must be conducted if electrical heating is being considered. Cooling mode: Chilled water is the preferred cooling medium for HVAC applications and should be used whenever practicable.

The direct expansion of refrigerant in coils is an acceptable method of cooling, particularly on small isolated plants, or where lower temperatures are needed for dehumidification or for cold room. This system, however, does not normally give close control. Direct expansion coils should only be used with extreme care on variable air volume systems (if speed driver available on compressors).

Heating Coils

The face velocity of air across heating coils should not exceed 2 m/s. Coils should be made of material suitable for applicable constraints. Drains shall be located outside the casing of the HVAC unit. Coils shall be removable.

Cooling Coils

Cooling coils have been identified as potential sources of microbial contamination; therefore, careful design is required to prevent water carryover and to ensure that drain pans do not retain water. Double tube, non-welded units are recommended. The face velocity of air across cooling coils should not exceed 2 m/s. Where necessary, stainless steel or plastic eliminator blades should be provided to prevent any moisture carryover. Where provided, these must be removable for cleaning.

Ductwork

For most applications, galvanised steel ductwork will be the most appropriate form of construction; however, stainless steel or plastic construction may be necessary where there is a higher risk of corrosion due to moisture or fumes (exhaust ducts usually). Where operating pressures above 2,000 Pa are necessary, fully welded construction is recommended. For contained ducts (e.g., exhaust duct before bag-in / bag-out filter), air tightness Class C shall be followed (EN 12237). For BSL-3, fully welded construction should be considered.

Generally ductwork should be constructed to an appropriate local standard, suitable for the maximum design pressure (positive or negative), such as those published by Sheet Metal and Air Conditioning Contractors' National Association (SMACNA) in the USA, Building and Engineering Services Association (B&ES) in the UK Where flexible connections are proposed these must be designed for the same pressure as the ductwork. Solid ducted connections are preferred for final connections to terminal HEPA filter housings. For applications where flexible connections to diffusers are used, these should be no longer than 500 mm and nominally straight. Special consideration must be given to fume extract ducts where these pass through fire barriers. Using fire dampers should be avoided where the loss of extraction could make a fire situation worse. An alternative design, such as the use of fire-rated ductwork, may be necessary in these cases. A thorough risk assessment must be conducted.

Simple Representation of HVAC system

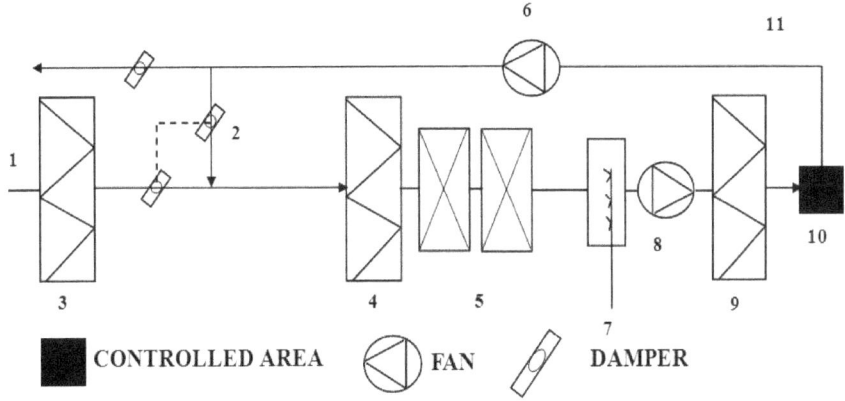

Position	Description
1	Fresh air intake (°C, %RH, flow rate)
2	Dampers
3	Filter creating a differential pressure
4	Filter creating a differential pressure
5	Control valves for cooling fluid
6	Exhaust fan
7	Steam flow rate
8	Supply fan
9	Filter creating a differential pressure
10	Controlled room/ area
11	Extraction

4.3. ISO Standards for Cleanrooms

ISO-14644: Cleanrooms and Associated Controlled Environments
ISO-14644-1 Classification of Air Cleanliness
ISO-14644-2 Cleanroom Testing for Compliance
ISO-14644-3 Methods for Evaluating & Measuring Cleanrooms & Associated Controlled Environments
ISO-14644-4 Cleanroom Design & Construction
ISO-14644-5 Cleanroom Operations
ISO-14644-6 Terms, Definitions & Units
ISO-14644-7 Enhanced Clean Devices
ISO-14644-8 Molecular Contamination

For Biocontamination control the following standards apply:

ISO-14698- 1 Biocontamination: Control General Principles
ISO-14698-2 Biocontamination: Evaluation & Interpretation of Data
ISO-14698 -3 Biocontamination: Methodology for Measuring Efficiency of Cleaning Inert

4.4. Temperature

Unless otherwise required by the regulatory, product and/or process driven specifications, HVAC system design should be based on user selected temperature(s) within the range defined in Table 2. A facility should meet stated temperature design conditions; however, the acceptability of the facility or operation depends on meeting the operating ranges.

4.5. Air Handling Units

All GMP Air-Handling Units (AHU) should have the capability of being custom designed and constructed to meet the more stringent operation and maintenance requirements for these areas. All GMP AHUs should be located in an internal plantroom to avoid risk of contamination during maintenance.

The frame shall be constructed from heavy gauge box section steel or robust aluminum structure and supported on a sectional steel channel. The AHU Casing shall be constructed of modular double skin panel which shall be of cold bridge free construction. Casing panels shall be manufactured from sheet steel, with galvanized inner skins and pre-painted outer skins. Panels shall be 60mm thick and filled with CFC free water-based PU insulation foam and shall be FM approved and meet NFPA fire rating. Casing shall have a U value of less than 0.55 W/m^2K. Enclosure panels shall be manufactured from 1.3 mm galvanized sheet steel and painted PVC coated finish.

All surfaces exposed to airstream shall be hot dipped galvanized or Type 304 stainless steel as indicated on the datasheets. Aluminum will be accepted in lieu of galvanized or stainless steel

An AHU with a cooling coil shall have a drift eliminator and stainless-steel drain trays of adequate size to collect water with a minimum depth of 40mm. The drain trays shall be extended past the coils to capture all moisture carryover from the coil by the airstream. All trays shall be inclined towards the drain connection and traps adequately sized to resist the positive or negative pressure at that point in the AHU.

4.6. Filtration

Air filters are the primary method to reduce contamination levels in an air stream and play a very important role achieving the clean room environment. It is not only the efficiency of the filter that is important to address, but also the energy consumption (the pressure drop during the entire operation).

The air change rates should be determined by the manufacturer and designer, taking into account the various critical parameters using a risk based approach with due consideration of capital and running costs and energy usage. Primarily the air change rate should be set to a level that will achieve the required clean area condition.

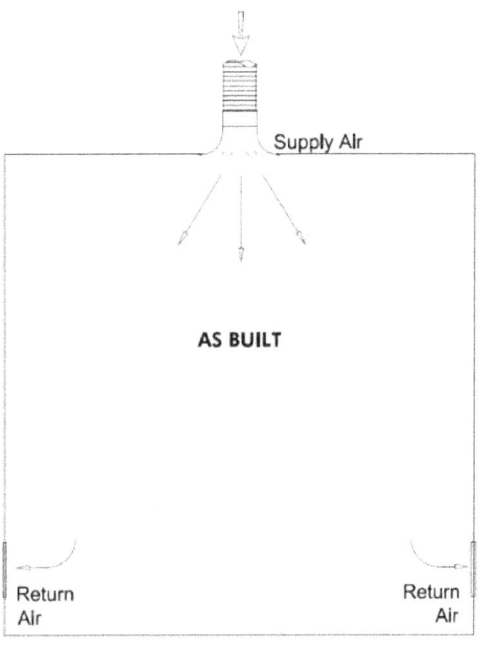

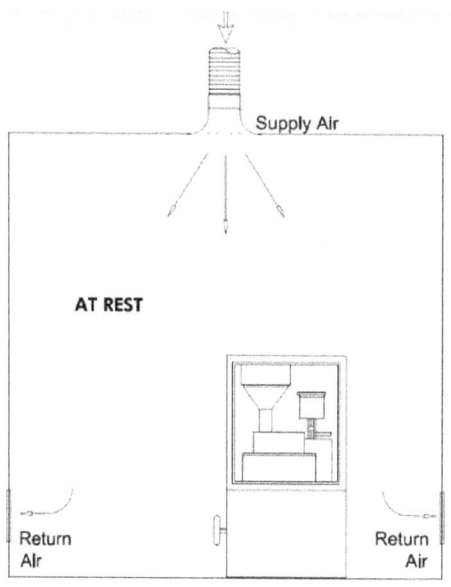

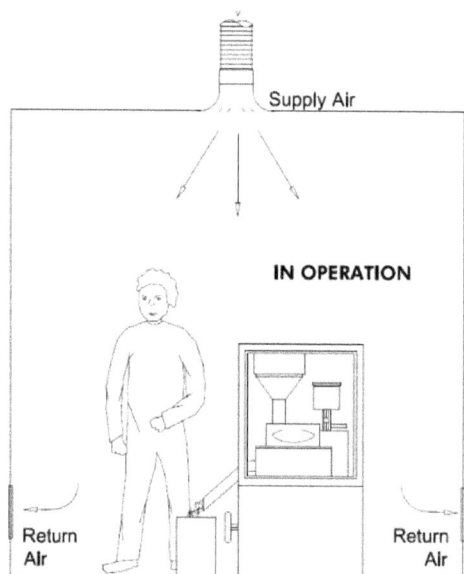

Room classification tests in the "as-built" condition should be carried out on the bare room, in the absence of any equipment or personnel. Room classification tests in the "at-rest" condition should be carried out with the equipment operating where relevant, but without any operators. Because of the amounts of dust usually generated in a solid dosage facility most clean area classifications are rated for the "at-rest" condition.

Room classification tests in the "operational" condition should be carried out during the normal production process with equipment operating, and the normal number of personnel present in the room. Generally a room that is tested for an "operational" condition should be able to be cleaned up to the "at-rest" clean area classification after a short clean-up time. The clean-up time should be determined through validation and is generally of the order of 20 minutes.

Materials and products should be protected from contamination and cross contamination during all stages of manufacture for cross contamination control.

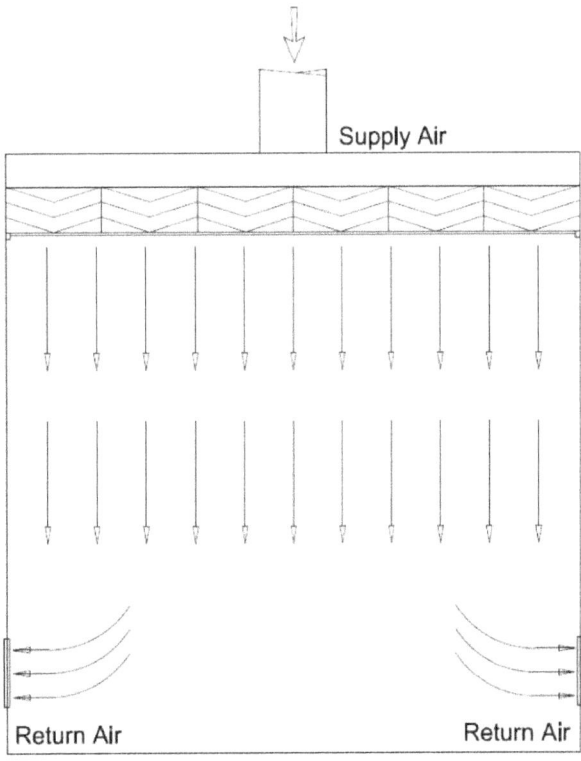

4.7. course/ PRE- Filtration

Pre/course filtration shall be located in the AHU just after the outside and return air streams enter the recirculation unit. Level 1 filtration is the lowest efficiency, lowest cost and is used to remove all large particles (3.0 microns and larger such as insects and vegetation) found in outside air. The intention is to keep the internal components, (coils, fans etc.) and the AHU internal surfaces clean over an extended period. They also act as a pre-filtration for the Level 2 filtration and extend their life. A minimum of EN G4 (MERV 7) filters are recommended for Level 1.

Filter face air velocities shall not exceed of 2.5m/s (450 fpm). At the AHU maximum air volume flow rate, the initial pressure drop (clean) across the filter shall not exceed 100 Pa and the final pressure drops (dirty) should not exceed 250 Pa (1.0" w.g.) for panel/pleat filters as guidance, before completing the LCC analysis.

4.8. Fine / SECONDARY Filtration

Secondary or Fine filtration is more expensive and should be located as the last component before discharge from the AHU. This is recommended to ensure any particles or other matter (mold) generated in the AHU is captured before discharge to the ductwork and also to extend the life of filters further downstream. EN F8 or F9 (MERV 14/15/16) filters are recommended for Level 2.

Filter face air velocities shall not exceed of 2.5m/s (450 fpm). At the AHU maximum air volume flow rate, the initial pressure drop (clean) across the filter shall not exceed 100 Pa and the final pressure drops (dirty) should not exceed 450 Pa (1.4" w.g.) for panel/pleat filters as guidance, before completing the LCC analysis.

Grade	Maximum permitted number of particles per m³ equal to or greater than the tabulated size			
	At rest		In operation	
	0.5 µm	5.0 µm	0.5 µm	5.0 µm
A	3 520	20	3 520	20
B	3 520	29	352 000	2 900
C	352 000	2 900	3 520 000	29 000
D	3 520 000	29 000	Not defined	Not defined

maximum permitted airborne particle concentration for each grade. Showing both "at rest" and "in operation" conditions. (EU V4 Annex 1). The EU guidance given for the maximum permitted number of particles in the "at rest" column corresponds approximately to the ISO classifications.

Room Air Classification (By Limits of Microbial Contamination)

The HVAC systems help maintain the viable (microbial) limits within a specific area. These limits are defined in Annex 1 of the EU GMP Guide as shown below:

Grade	Recommended limits for microbial contamination (a)			
	air sample cfu/m³	settle plates (diameter 90 mm) cfu/4 hours (b)	contact plates (diameter 55 mm) cfu/plate	glove print 5 fingers cfu/glove
A	< 1	< 1	< 1	< 1
B	10	5	5	5
C	100	50	25	-
D	200	100	50	-

4.9. Compliance Tests for GMP Zones

4.9.1. Particle count test
Test covers verification of cleanliness. Dust particle counts to be carried out and result printed. The number of readings and positions of tests should be defined in accordance with ISO 14644-1 Annex B5

4.9.2. Filter Leakage Tests
To verify filter integrity. Filter penetration tests to be carried out by a competent person to demonstrate filter media, filter seal and filter frame integrity. Only required on HEPA filters. Refer to ISO 14644-3 Annex B6

4.9.3. Containment Leakage Test
To verify absence of cross-contamination. Demonstrate that contaminant is maintained within a room by means of:
• airflow direction smoke tests
• room air pressures.
Refer to ISO 14644-3 Annex B4

4.9.4. Air Pressure Differential
This test is used to verify 'non cross-contamination'- positive air pressure pushes out particles from clean zone. Log of pressure differential readings to be produced or critical plants should be logged daily, preferably continuously. A 15 Pa pressure differential between different zones is recommended. Refer to ISO 14644-3 Annex B5

4.9.5. Air Flow Volume
To verify air change rates. Airflow readings for supply air and return air grilles to be measured and air change rates to be calculated. Refer to ISO 14644-3 Annex B13

4.9.6. Air flow velocity
To verify unidirectional flow or containment conditions. Air velocities for containment systems and unidirectional flow protection systems to be measured. Refer to ISO 14644-3 Annex B4

4.9.7. RECOVERY
To verify clean-up time. Test to establish time that a cleanroom takes to recover from a contaminated condition to the specified cleanroom condition. Should not take more than 15 minutes. Refer to ISO 14644-3 Annex B13

4.10. Air Flow Visualisation
To verify required airflow patterns. Tests to demonstrate air flows:
- from clean to dirty areas
- do not cause cross-contamination
- uniformly from unidirectional airflow units

Demonstrated by actual or video-taped smoke tests. Refer to ISO 14644-3 Annex B7

4.11. Clean Room Design Considerations

4.11.1. Seasonal Variations
All locations on earth except latitudes near the equator experience seasonal temperature changes. The changes are a consequence of Earth's orbital motion about the sun, coupled with the tilt of its axis of rotation with respect to its orbital plane. Design criteria should be based on published temperature data. The HVAC system design should consider the following:

Standard Operating Conditions: These are climatic conditions against which the systems must be designed to operate, control, and maintain required conditions. (These may be based on published data, which are only exceeded 2.5% or 1% of the time).

Extreme Operating Conditions: These are climatic conditions against which the systems must be designed to operate, without manual intervention, and without damage to the systems or the facility. Based on product / process risk assessments, extreme or standard conditions shall be used for HVAC design for dedicated areas.

Location
Based on the building layout, footprint and design intent, a suitable and adequate space must be identified for HVAC location. This must include provision of chilled water, heating systems, ducts and drainage. HVAC plants must be accommodated in designated HVAC plant rooms or interstitial areas.

Thermal Load

Thermal load can be defined as the amount of heat energy to be removed from an inner environment by equipment (HVAC) used to maintain that environment at the design temperature when worst case external temperature(s) are being experienced. The thermal load requirement should be calculated for the following:

- Max summer conditions
- Minimum winter conditions
- High rainfall
- Standard operation
- Extreme operating conditions

4.11.2. Dust, Vapour, or Fume Control

Highlight areas requiring dust, vapour, gas and/or fume control on the room data sheet. These areas must be controlled to remove the possibility of product contamination and to ensure the safety of the operator and environment. Areas requiring 100% fresh air or extraction to atmosphere may require greater airflow or other measures within the room to maintain environmental conditions.

In order to meet the appropriate level of cleanliness, HVAC systems require sufficient filtration to provide "clean" air to prevent contamination of the product. Pre-filters and main filters are normally suitable for most operations; however, HEPA filters are required to prevent particulate or microbial contamination for higher-classification areas

<u>Air Change Rates</u>

The air change rates for each room must be calculated to be sufficient for clean-up to achieve specified particulate conditions "at rest" in static conditions after a maximum of 20 minutes from completion of operations. The actual air change rate must be chosen to satisfy the most stringent requirements including GMP, GLP, heat gain, ventilation requirements and/or occupancy, including an appropriate safety factor.

The air change rate must be optimised for energy savings; however, specific attention must be paid to air locks where a greater air change rate must be applied. Air changes can be reduced (e.g. setback modes) in some circumstances ("at rest" mode, with no production activity and no personnel interventions).

<u>Room Exhaust</u>

Where there is a risk of active compounds being present in extracted air, filters should be fitted, preferably in the room, to prevent contamination of ductwork and the environment. The filters must be selected based on the particle size distribution of the products to be handled.

<u>Dust Extraction and Collection</u>

It is essential to capture dust as close as possible to the point of generation without affecting the process. In most cases dust capture should be within 100mm from the point of release. Air velocity is the key parameter in dust capture.
Pharmaceutical and chemical applications have specific collection requirements as any dust build-up in the system is likely to be of a pharmacologically active nature, sensitising, toxic and/or corrosive. It is vital to maintain transport velocities and minimise any potential for cross contamination.

A typical system should have a minimum transport velocity of 18 m/s, but this may need to be higher if heavy particles are to be collected. This velocity must be maintained throughout the system to prevent dust from dropping out in the ducts.

The dust collection must be configured with the hazardous nature of the dust in mind. A clearly defined disposal procedure for the collected dust (e.g. bag-in / bag-out system for filter and dust bin) needs to be understood at the design stage. HVAC unit shall meet EN 1886 and EN 13053 requirements.

Fans

Certified performance curves are required to verify correct fan operation. Fans that may be subjected to high temperatures, humidity, corrosive fumes or other hazardous atmospheres should be constructed using non-reactive, non-corrosive, suitable and approved materials (such as epoxy painting). Whenever H2O2 or other disinfection application is planned, material compatibility certificates shall be supplied by the vendor.

Fans must be selected to supply the design volume, taking into account the assumption that filters are half clogged, except for the terminal filter which shall be considered to be fully clogged according to EN 13053. If the terminal filter is HEPA, clogging shall be considered according to EN 1822 and the target volume is 80% of the given maximum clogged specified value.

Filtration

Face-fitting filters shall be used in all cases, as slide-in filter elements never give a good seal. The installation must be such that the airflow pushes the filter against the seal. The face velocity across the filter section shall not exceed 2 m/s. For ventilation and air conditioning applications, two minimum filtration stages are required. For certain applications, return air filtration will be required to contain highly active materials (e.g. viruses or potent compounds). Normally, these filters should be changed from the room side. However, since those filters must be integrity tested, it is recommended to place one filter in the main return duct before the exhaust fan and design return duct network, in order to ensure tightness of the duct between the room and the filter (bag-in / bag-out filter change systems should be provided for BSL-3 areas). In case of live biological agent biocontainment, decontamination up to the filter must be proven. The grade of filter and technical solution must be selected based on the product particle size distribution and occupational exposure band (OEB) level.

5. Utility Gases & Water

5.1. Introduction

The key utilities involved for cleaning include utilities such as water, compressed gases (air, nitrogen etc.) and the heating and cooling of process equipment. Water quality can impact the effectiveness of pre-rinsing, washing, and final rinsing. Therefore, both the water temperature and quality need to be tightly controlled and monitored. Gases are typically used in order to blowdown or blowout remaining fluids or they are used as a drying step.

The term "clean utilities" in the life science industry refers to utilities that have to fulfil quality regulatory requirements. The basis of these requirements is due to the application of the utilities in the production of products or if water (e.g. water for injection) is used in the final product, or cleaning or processing where product contact may occur.

The most common utility is water, which can be supplied in different pharmaceutical grades of purity. Purified water (PW or PUW), Highly Purified Water (HPW) and Water-for-Injection (WFI) are the most common. Water quality specifications can be found in the pharmacopeias, e.g. the US Pharmacopeia. Other clean utilities can also include clean compressed air, clean gasses (e.g. nitrogen, argon and oxygen), and clean steam.

5.2. CLEAN STEAM

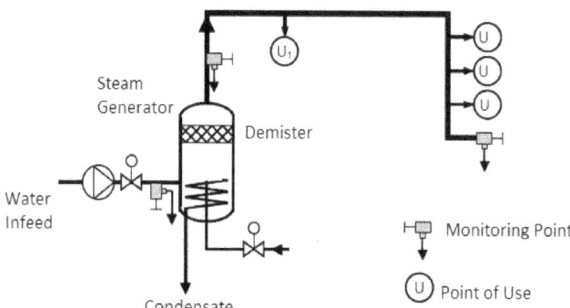

Clean Steam Generation Piping and Instrumentation

Pure steam is used in pharma and biotech for sterile application, for autoclave sterilisation etc. Distribution piping of clean steam is a critical aspect. Improper sizing of pipes may impact the production process and lead to a loss of time during sterilisation. Clean steam, also referred to as "pure steam", and gases used in manufacturing operations must be of a quality suitable for their intended purpose. The intended use of clean steam and gases must be understood in order to determine any risks to the patient or product. For example, gases that end up being part of the product must fulfil the regulatory requirements. Preventative maintenance and ongoing monitoring must be implemented for clean steam systems.

Water systems for purified water, de-ionised water and Water-for-Injection (WFI) must provide a consistent and reproducible output. Where there is moisture, there is always a risk of microbial contamination. Therefore, the design of water systems should mitigate against such risks. Good engineering practices such as using circulation loops, no dead legs and polished-surface finishes all work to provide an effective and safe system. The design should also take into account ease of sampling at the point of use. The removal of endotoxins is a requirement for WFI. Ongoing sampling monitoring the quality of water is particularly important where water systems are concerned. Procedures should be in place to ensure that effective monitoring and testing is maintained. Action limits and acceptance criteria should be clearly documented in approved SOPs or the equivalent. Failure to meet limits or acceptance criteria should initiate an investigation.

Overview of Clean Steam Generation

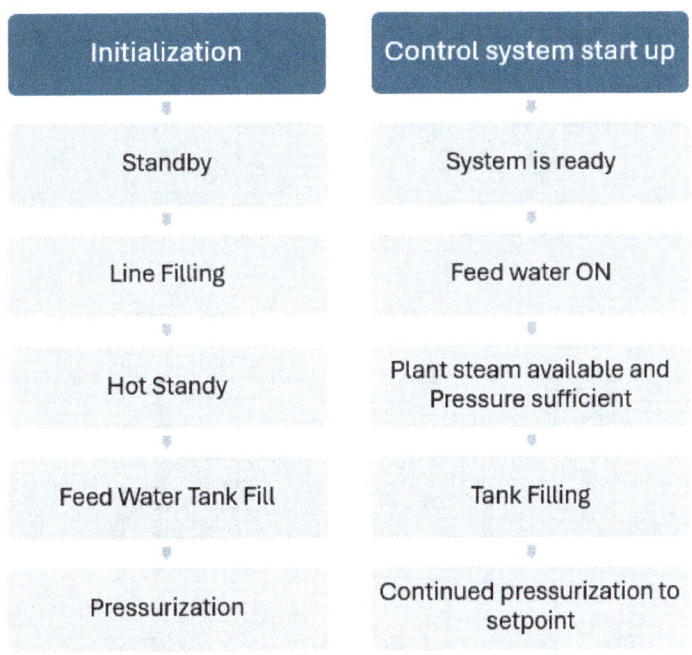

OQ Testing

Operational qualification or OQ is a formal validation activity, and as such should be completed per an approved protocol. The purpose of OQ testing is to confirm the operational and functionality of the clean steam system. This should demonstrate that all critical aspects of a URS are fulfilled. OQ verifications include:

- Testing of temperatures and operating pressures
- Capacity testing (under load)
- Steam trap operation
- Verification of automated functions and alarms
- Check of automation systems
- Correct function of valves and sampling points

PQ Testing

Due to the high operating temperatures and the associated lethality, clean steam systems are resistant to microbiological contamination.

Issues that arise can normally be attributed to equipment failures with the steam generator or contaminated water being supplied to the system. Bacterial endotoxin testing is used to monitor clean steam systems for both PQ purposes and throughout the life cycle of the equipment operation. Steam is condensed, sampled and tested. The condensate should meet WFI specifications with the exception of viable total aerobic count. Clean steam PQs are commonly completed using a three-phase approach to testing. The first phase ensures the system consistently operates within the required ranges and the steam provided meets the acceptance criteria. Typically, phase one bacterial endotoxin testing and physio-chemical testing is completed over a two-week period. For phase two, the same frequency and type of testing may be applied for an additional two weeks. After phase two testing, the system may be available for general use if allowed for within internal company procedures. Phase two testing at PQ should also provide a report with all results documented and reviewed. Phase three of PQ is intended to demonstrate the effective and consistent operation of the system over a longer term (approx. 12 months). Sampling is typically performed weekly.

Schematic representation (simplified) of clean steam system

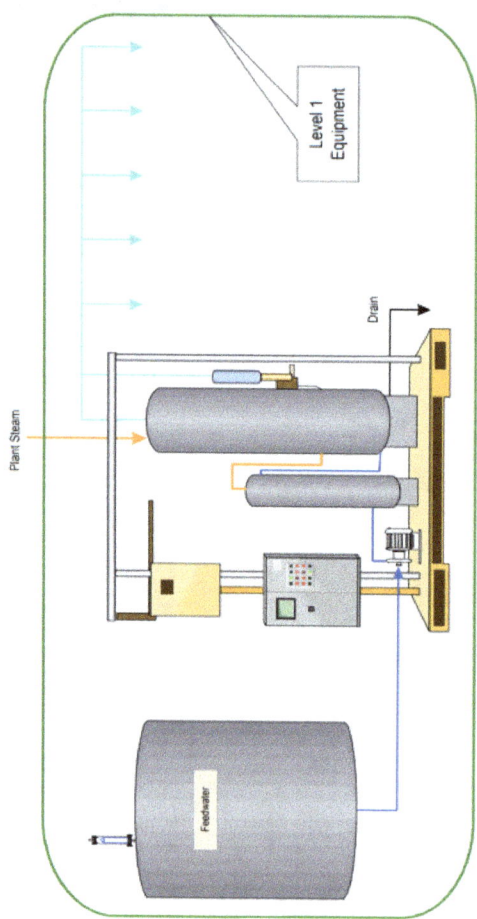

Further Reading on Clean Steam
- PIC/S PI009-3 – Pharmaceutical Inspection Co-Operation Scheme - Inspection of Utilities
- EN 285 – European Standard - Sterilisation, Steam Sterilisers, Large Sterilisers
- USP <1231> – United States Pharmacopeia - <1231> "Water for Pharmaceutical Purposes"

- USP– United States Pharmacopeia - Monograph "Pure Steam"
- EN 285 – European Standard - Sterilisation, Steam Sterilisers, Large Sterilisers

5.3. RO Water, DI water and Water for Injection

5.3.1. Water Systems
Water supply and the associated water systems in biotechnology and pharmaceutical plants are often critical utilities and therefore, critical to quality and safety or product. Purified water is commonly used to clean equipment and vessels, to cool or heat processing pipes and systems, in sterilize products or components via moist heat sterilsiation or indeed are used to the formulation of producing the finished product (e.g. water-for-injection). Various grades of water service a particular purpose. Some common types include:

- RO water
- DI Water
- Purified water
- Water-for injection

Reverse Osmosis, RO water and Deionized, DI water are both types of purified water. They are however, produced via different processes and therefore have different characteristics. RO water is produced by forcing water through a semi-permeable membrane. Water molecules can pass through the membrane material while larger molecules are due to their size are prevented from traversing the membrane. The RO process removes many impurities, but some dissolved solids or impurities or contaminants may still be present in the water after the RO process. DI water is created by passing water through an ion exchange membrane or material that that removes charged ions by the deionization process. Therefore, DI water seen as a more purer water that is suitable for applications in pharmaceutical manufacturing where no impurities or contaminations are desired. Critical Process Parameters for a water system include:

- Pressure
- pH
- Conductivity Level
- TOC
- Flow rate
- Temperature
- Resistivity

5.4. High Purity Water System Design
Design requirements need to be based on the intended use of the high purity water output. The use or application of the high purity water informs the design teams of the type, size, performance and specifications of the system. For example, most inhalation and ophthalmic products, topical products and oral products use purified water as part of the final formulation. Additionally, a higher specification of purified water is used in parenteral products as Water for Injection. System design also needs to take into account the temperature of the system. Hot systems are beneficial as they can be configured and validated as self-sanitizing systems ((65DegC – 80DegC).

In addition to intended use and temperature, deciding if a system will be circulating or one-way. The selection may be informed by the usage requirements of the manufacturing process and intended use. A risk assessment is important to identifying potential risks and mitigations associated with water systems and if different products and activities require different levels of purification. One-way systems can be described as dead leg systems and may be at greater risk of contamination and out of specifications.

Water for injection is achieved either by Distillation or Reverse Osmosis (RO) filtration are the only acceptable methods listed in the USP for producing Water for Injection. Injectables require extremely pure and sterile water that does not contain endotoxins and complies with USP standards and applicable regulations.

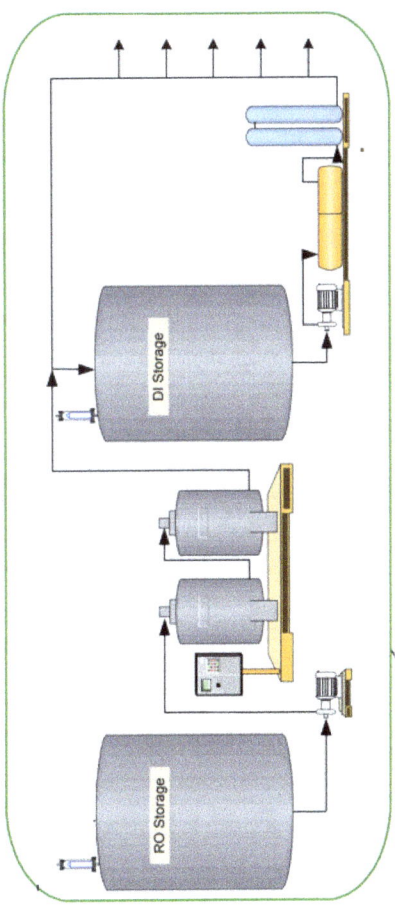

The CQAs and CPPs are routinely monitored through the calibrated monitoring system which ensures any equipment failures would be detected.
- Pressure
- pH
- Conductivity Level
- TOC
- Flow rate
- Temperature
- Resistivity

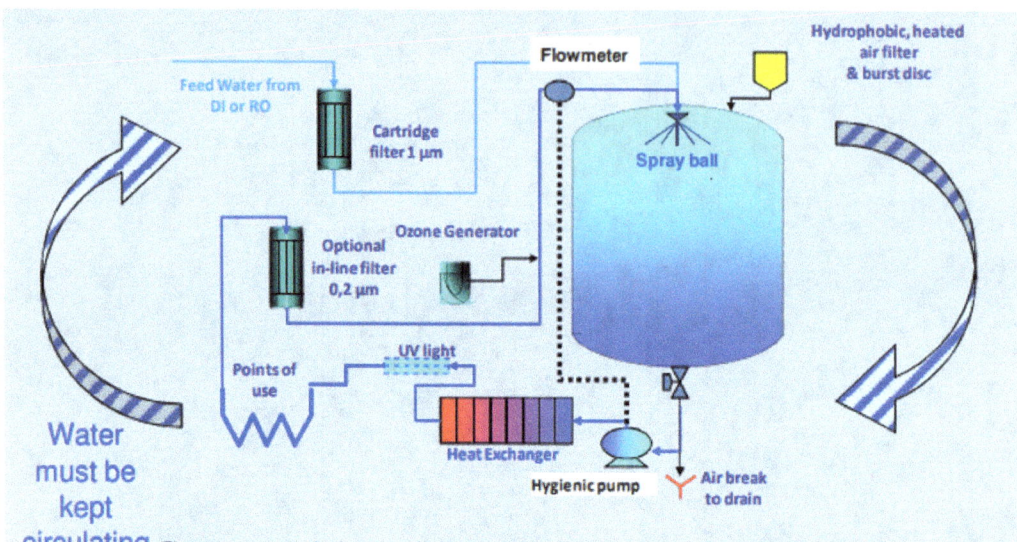

Flowmeter controls the speed of the pump, to guarantee sufficient flow speed of water in the loop (**turbulent flow**, to prevent build-up of biofilms)

5.5. Water for Injection

WFI is sterile and pyrogen-free water containing no less than 10 CFU/100ml (Colony Forming Units) with a sample size of between 100 and 300 ml and an endotoxin level < 0.25 EU/ml. The use of WFI is two-fold. Firstly, it can be used for critical processing steps such as washing and rinsing. It can also be used in injectable products. WFI is a key raw material for sterile intravenous and intradermal products. WFI is produced by a Multi-Column Distillation Plant (MCDP) and must meet the microbial requirements of regulated bodies.

WFI generation, storage and distribution

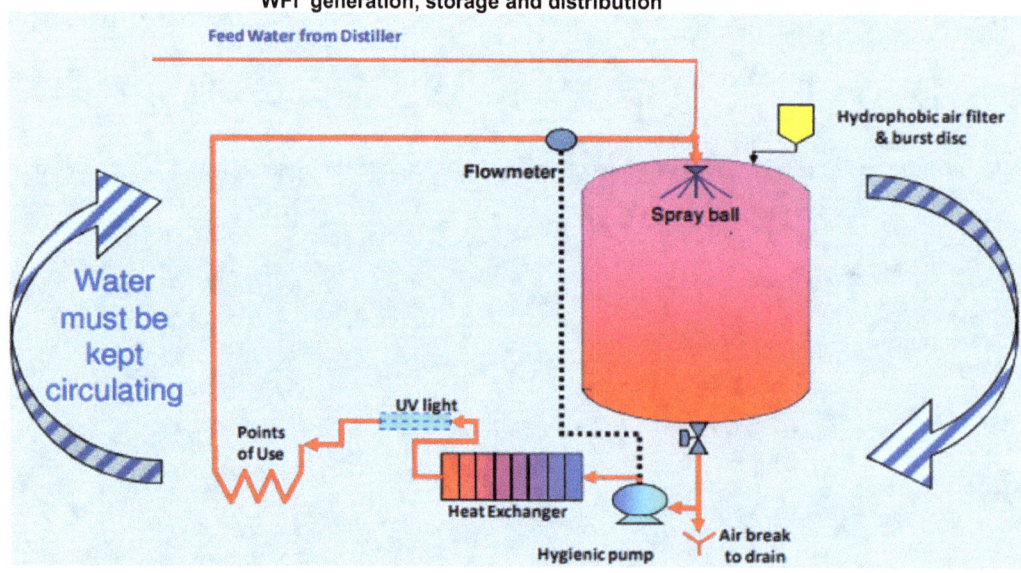

This schematic illustrates the requirement for WFI circulation to be maintained continuously via a hygenic pump. Heat exchangers and UV light ensure microbial protection. Multiple points of use can be included within the distribution system.

Piping in WFI systems usually consist of a high polished stainless steel. In a few cases, manufacturers have begun to utilize PVDF (polyvinylidene fluoride) piping. It is purported that this piping can tolerate heat with no extractables being leached. A major problem with PVDF tubing is that it requires considerable support. When this tubing is heated, it tends to sag and may stress the weld (fusion) connection and result in leakage. Additionally, initially at least, fluoride levels are high.

This piping is of benefit in product delivery systems where low level metal contamination may accelerate the degradation of drug product, such as in the Biotech industry. One common problem with piping is that of "dead-legs".
With colder systems (65 - 75oC), any drops or unused portion of any length of piping has the potential for the formation of a biofilm and should be eliminated if possible or have special sanitizing procedures. There should be no threaded fittings in a pharmaceutical water system. All pipe joints must utilize sanitary fittings or be butt welded. Sanitary fittings will usually be used where the piping meets valves, tanks and other equipment that must be removed for maintenance or replacement.
A companys procedures for sanitization, as well as the actual piping, should be reviewed and evaluated during the inspection.

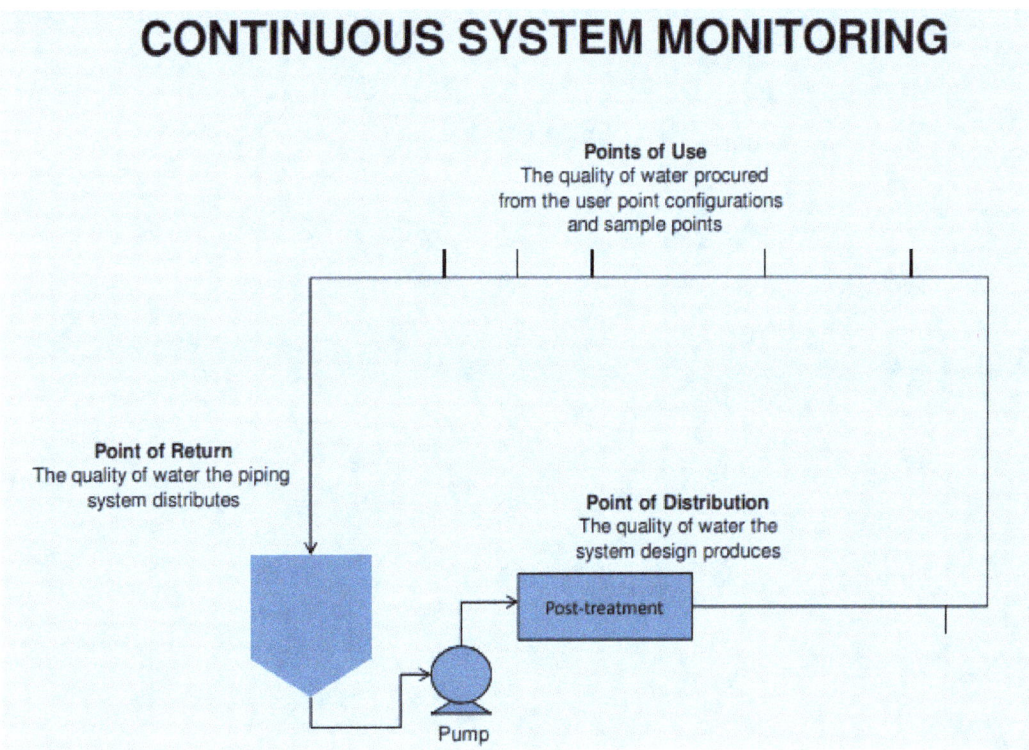

The cleaning of equipment, vessels and process piping is a critical activity. Any residue from a previous production batch needs to be removed in order to avoid cross-contamination. Clean in Place and Sterilize in Place skids are often utilised to allow efficient switchover between batches and/or products. Where possible, manual cleaning should be avoided unless essential due to the design of a system or particular location or configuration.

WFI relies of feedwater as an input to producing WFI distillate. Feedwater endotoxin and microbial levels can vary from season to season. A water system still should be designed to operate within a range of anticipated environmental extremes due to these seasonal effects. Obviously, the only way to know the extremes is to periodically monitor feedwater. If the feedwater is from a municipal water system, reports from the municipality testing can be used in lieu of in-house testing. Pretreatment of feedwater is recommended by most manufacturers of distillation equipment and is definitely required for RO units. The incoming feedwater quality may fluctuate during the life of the system depending upon seasonal variations and other external factors beyond the control of the pharmaceutical facility.

A potential risk with WFI systems is the failure to pretreat feedwater in order to reduce levels of endotoxins. Still fabricators may only guarantee a 2.5 log to 3 log reduction in the endotoxin content. Variation in the quality of feedwater can then lead to unacceptable levels of endotoxins in the WFI distillate. Pretreatment systems for the stills include the use of deionization in combination with ultrafiltration, RO or distillation. Conductivity meters are used on water systems to monitor chemical quality and have no meaning regarding microbiological quality.

5.6. Microbial Limits

Organisms exist in a water system either as free floating in the water or attached to the walls of the pipework, tanks and surfaces of valves. When they are attached to the walls they are known as biofilm, which continuously slough off organisms. **Limits are established by the manufacturer and represent action limits.** For WFI that standard action limit applied is less than 10 CFU/100ml. The purpose of establishing any action limit or level is to assure that the water system is under control.

A sample size of between 100 - 300 mL is preferred when sampling Water for Injection systems. However, sample volumes less than 100 mL are unacceptable. The sampling method may introduce contamination due to the fact that it is typically performed in non-sterile areas and is not truly aseptic, Therefore, occasional low level counts due to poor sampling may occur. Instances where action limits are exceeded requires the manufacturer to root cause the occurrences and take the necessary actions to correct the problem. An assessment on the impact of the breaches on the products must be documented also. WFI is also monitored for endotoxins which is testing using the LAL endotoxin test.

5.7. Purified Water Systems

The microbial requirements for purified water systems is not as clear as water for injection. But the expectation is that purified water is free from objectionable organisms. These are defined as any organisms that can cause infections when the drug product is used as directed or any organism capable of growth in the drug product. FDA requirements specify that any action limit over 100 CFU/mL for a purified water system is unacceptable. Action limits should be informed by the requirement and specification of the overall purified water system and what further processing of the finished product occurs. Noting that Purified water used to manufacture drug products by cold processing should be free of objectionable organisms. Contamination is not uniformly distributed in a water system. A given sample may not be representative of the type and level of contamination throughout the system as a whole. A count of 10 CFU/mL in one sample and 100 or 1000 CFU/mL in a subsequent sample would not be unrealistic. Therefore the cadence, number of samples and locations of sampling needs to be determined based on risk and expertise with trending to determine any emerging issues or contamination.

6. Compressed Air- Generation, storage and distribution

6.1. Introduction

Compressed air is used for valve actuation, instrument air and process air to name but a few applications. Only the point-of-use filtration and the gas quality instrumentation should be classified as level 1. When flow or pressure is a CPP, the measurement/monitoring should be performed by the system into which the gas is flowing. Additionally, the CQAs and CPPs should be routinely monitored through the calibrated monitoring system. For compressed air, the potential CPPs are listed below. For the physical system being evaluated, the use and the application of the compressed air will determine which (if not all) CPPs are needed to ensure the system produces product of the desired quality.

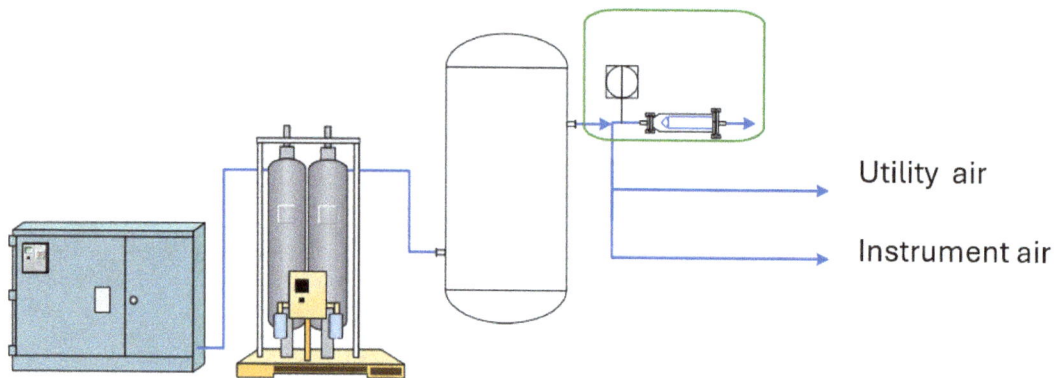

- Hydrocarbons
- Moisture
- Particulates
- Temperature

It is important that each point of use has appropriate sterile filters in place. If the filter is not placed directly at the point of use, control and counter measures should be implemented to address any risk of contamination downstream of the filter. Compressed air for bio-pharmaceutical use must be generated using oil free compressors with appropriate temperature controls in place.

Requirement	Clean Compressed Air (impacts product quality)	Sterile Compressed Air (impacts sterility of product)
Oil content	*Not great than 0.1mg/m^3 (ISO 8573-1 Class 2)	
Microbiological requirement	Meets requirements of the environmental zone served (e.g. ISO 5)	Sterile
Filtration requirement	Minimum 0.45μm membrane filter	0.2μm membrane filter

6.2. Compressed Air Design Requirements

Compressed Air generation systems are required to address the following components in order to produce compressed air that complies to ISO 8573-1 requirements.

Class	ISO 8573-1					Viable particle counts by Air sampling Method
	Solid Particulate			Water content	Oil content	
	Maximum no. of particle per m³			Vapor pressure Dew point	Total oil mg/ m³	
	0.1-0.5 μ	0.5-1μ	1-5μ			
0	As specified by the user / supplier (≤ class 1)					
1	100	1	0	-70°C	0.01	100 CFU/m³
2	100,000	1000	10	-40°C	0.1	
3	-	10,000	500	-20°C	1	

6.3. Design Element: Inlet Air filters

Purpose: Inlet air filter is required in order to remove particles from the atmospheric air entering the compressor system.

6.3.1. Design Element: Air Compressor

Purpose: The compressor acts to compress the air into a small volume, and increases the pressure.

6.3.2. Design Element: Inter Cooler

Purpose: The inter cooler lowers the temperature of hot and wet air leaving from first stage air compressor by removing water as condensate. The air then enters the second stage compression to achieve desired pressure and quality.

6.3.3. Design Element: After Cooler

Purpose: The after cooler lowers the temperature of hot and wet air leaving from second stage air compressor by removing water as condensate.

6.3.4. Design Element : Dryer

Purpose: Dryer function is normally inbuilt in compressor and is able to eliminate any remaining moisture in the compressed air leaving from after cooler.

6.4. Design Requirements

Design elements are then translated into specific and detailed design requirements. The design requirements of compressed air generation & distribution system are specified below.

1. **General Requirements**
    i. Capacity: of the air generation system : capacity must be calculated by determining the usage requirements of the equipment and or facilities that requires compressed air

ii. Storage Vessel Capacity: to be specified
iii. Outlet Pressure10,00 m³/min: per equipment/ facility requirements e.g Maximum 10,00 m³/min @ 55,4 Hz Frequency, Maximum 4,81 m³/min @ 31,2 Hz Frequency

2. **Compressed Air Generation system**
   i. Inlet Filter: meets ISO requirements
   ii. Air Compressor Capacity : specified based on demand/requirements
   iii. Make: Preferred make/model or similar if required
   iv. Dryer: Inbuilt compressor unit and Heat of compression type dryer or better to produce Dew point -20°C or better as per ISO 8573
   Discharge Pressure: as recommended
   Online Dew point Sensor: with display device
   Inline Filters: 10μ, 5μ, 1μ shall be installed just after Air compressor out

3. **Compressed Air Distribution system**
   i. Air receiver tank requirement Capacity: based on demand estimates
   ii. Quantity: as above
   iii. MOC: Material of construction e.g. SS304
   iv. Drain Valve: Auto and manual type
   v. Pressure Gauge:
   vi. Safety Valve: As per supplier design
   vii. Distribution Line   MOC: SS 304
   viii. User Point /User Valve: Screw ended and manual type, MOC-SS 316
   ix. Filter at user end (Process Area): 0.2 μ
   x. Quantity (Approx.): per requirements
   xi. Compressed Air Generation System: Emergency ON/OFF switch
   xii. Safety valve on Air Receiver tank

4. **Utility Requirement**
   i. Available utilities Power - 3Ø (Phase), 380 – 440 VAC
   ii. Documentation & Drawings
   iii. Turnover packet, Design qualification (DQ), FAT, Installation qualification (IQ), Operational qualification (OQ) protocol
   - General Assembly drawing of Air compressor
   - Test certificate required for filter of compressor
   - Test Certificate of the compressed discharge air class
   - Instrument calibration certificates including copies of certificates of test equipment used in calibration
   - Critical / recommended Spare part lists with quotation

5. **Compressed Air Distribution System Documentation**
   i. Technical Catalogue of all Bought outs of components/instruments if installed.
   ii. Air receiver tank and Distribution piping MOC certificate
   iii. Leak test report of supply line and header
   iv. Cartridge filter certificates

6.5. Design Qualification

So far, in relation to compressed air generation, distribution and storage we have summarized the design elements and design requirements. (Previous sections). Both the design elements and requirements are inputs to the Design Qualification. At this point in a project an approved User requirements specification should also be available. DQ is an evaluation of the design elements and design requirements that the URS and Vendor specifications. Note: Vendor specifications are often documented in a Functional design specification (FDS) which in simple terms is an 'answer' to each of the requirements specified in the URS.

6.6. DQ Evaluation

In this section the design requirements are benchmarked against the URS and Vendor responses or specifications. The Description is based on the design element and design requirement. The URS requirements are assumed to be already approved in the separate document. The vendor specification is the response of the vendor and can be a specific document that is created or alternatively if oof-the-shelf it may be a operating manual of similar document.

Description: Capacity
- User Requirements Specification: Generation of 1400 CFM with outlet pressure of 6-8 kg/cm2.
- Vendor Specification: Generation of 1600 cfm with outlet pressure of 6-8 kg/cm2.
- Verification: While the vendor specified system has a higher capacity, this is acceptable.

Description: Inlet air filtration
- User Requirements Specification: 3 microns with 99% efficiency
- Vendor Specification: 3 microns with 99% efficiency
- Verification: Requirement is met by design and vendor.

Description: Compressed air generation
- User Requirements Specification: Screw, non-lubricated oil free, air cooled.
- Vendor Specification: screw, non-lubricated oil free, air cooled.
- Verification: Require met by vendor

Description: Inter cooler
- User Requirements Specification: Air or water cooled
- Vendor Specification: air cooled type
- Verification: Air cooled type is acceptable.

Description: After cooler
- User Requirements Specification: Air or water cooled
- Vendor Specification: air cooled type
- Verification: Air cooled type is acceptable.

Description: Dryer
- User Requirements Specification: Must be inbuilt to the compressor unit with dryer to produce dew point -20DegC or better per ISO 8573
- Vendor Specification: Generation of 1600 cfm with outlet pressure of 6-8 kg/cm2.
- Verification: as specified above

This process is then replicated for the remaining user requirements. A successful DQ review will ensure all design aspects are review with acceptable vendor responses to the user requirements and design intent.

7. Clean steam

7.1. Introduction

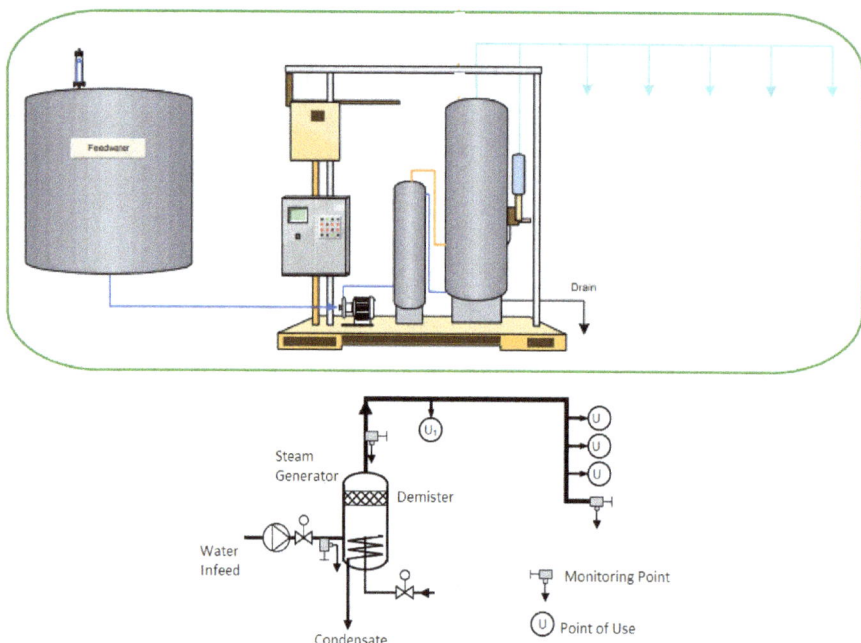

Clean Steam Generation, Piping and Instrumentation

Pure Clean Steam is used in for different functions in sterile manufacturing or used in autoclaving- moist heat sterilization. Distribution piping of clean steam is a critical aspect. Improper sizing of pipes may impact the production process and lead to loss of time during sterilisation. Clean steam used in manufacturing operations must be of a quality suitable for their intended purpose. The intended use of clean steam and gases must be understood in order to determine any risks to the patient or product. For example, gases that end up being part of the product must fulfil the regulatory requirements. Preventative maintenance and on-going monitoring must be implemented for clean steam systems.

- Routine inspection and maintenance
- Frequency of filter change
- Frequency of the sterilisation for the gas distribution system, if applicable
- Frequency for integrity testing of the sterile filter

Water systems for purified water, de-ionised water and water-for-injection (WFI) must provide a consistent and reproducible output. Where there is moisture, there is always a risk of microbial contamination. Therefore, the design of water systems should mitigate against such risks. Good engineering practices such as using circulation loops, no dead legs and polished surface finishes all work to provide an effective and safe system. The design should also take into account ease of sampling at the point of use. The removal of endotoxins is a requirement for WFI. On-going sampling to monitor the quality of water is particularly important where water systems are concerned. Procedures should be in place to ensure effective monitoring and testing is maintained. Action limits and acceptance criteria should be clearly documented in approved SOPs or equivalent. Failure to meet limits or acceptance criteria should initiate an investigation. The potential CPPs are listed below for clean steam systems:

- Conductivity
- Flow Level
- Pressure
- Resistivity
- Temperature

8. Facilities Monitoring

8.1. Monitoring Program for Personal
Personnel can significantly affect the quality of the environment in which the sterile product is processed. A vigilant and responsive personnel monitoring program should be established. Monitoring should be accomplished by obtaining surface samples of each operator's gloves on a daily basis, or in association with each lot.
This sampling should be accompanied by an appropriate sampling frequency for other strategically selected locations of the gown.

The quality control unit should establish a more comprehensive monitoring program for operators involved in operations which are especially labor intensive (i.e., those requiring repeated or complex aseptic manipulations). Asepsis is fundamental to an aseptic processing operation. An ongoing goal for manufacturing personnel in the aseptic processing room is to maintain contamination-free gloves and gowns throughout operations. Sanitizing gloves just prior to sampling is inappropriate because it can prevent recovery of microorganisms that were present during an aseptic manipulation. When operators exceed established levels or show an adverse trend, an investigation should be conducted promptly. Follow-up actions can include increased sampling, increased observation, retraining, gowning requalification, and in certain instances, reassignment of the individual to operations outside of the aseptic manufacturing area.

8.2. Elements of Monitoring
Control, maintenance, and system monitoring of cleanrooms must be conducted in accordance with defined standard operating procedures and includes but is not limited to the following activities:
- BMS/FMS management, including alert and/or alarm conditions
- Energy system management, as applicable
- Preventative maintenance
- Out of tolerance/event notification
- Cleanroom stop/restart management
- Periodic evaluation of cleanroom environments, including, but not limited to, HEPA filtration evaluation, hood certifications, and differential pressure testing

8.3. Building Management Systems
A Building Management System (BMS) functions as an automated control system designed to oversee various aspects of a building and its facilities, including heating, ventilation, air conditioning, security, fire protection systems, and more. This system comprises numerous Input/Output subsystems, controllers, servers, and workstations interconnected over a control network. Its primary purpose is to regulate, monitor, alert, and track equipment operations.

BMS systems are alternatively known as Facilities Management/Monitoring Systems (FMS), Energy Management Systems (EMS), Building Automation Systems (BAS), or similar terms.

Environmental Monitoring Systems (EMS) operate similarly as automated control systems. They consist of Input/Output subsystems, controllers, servers, and workstations linked over a control network. Their main function is to monitor, alert, and track critical environmental process parameters such as temperature, humidity, differential pressure, conductivity, and the status of coolers/refrigerators, among others. Here's a suggested classification of BMS and EMS systems based on their intended use.

Building Management System (BMS)	
System Classification	GxP
Data Usage	Data not used for GxP impacting decisions.

	Engineering use only
Monitoring	No Critical process parameters are monitored by the system
Controls	No GxP equipment
System Boundaries	Up to the point of use of the system or equipment
Validation	Not required

Environmental Monitoring System (BMS)	
System Classification	Non GxP
Data Usage	Data may be used to make quality decisions and product release decisions. Data is used to determine compliance.
Monitoring	Critical process parameters are monitored by the system
Controls	Critical alarm limits are controlled
System Boundaries	From the point of use
Validation	Not required

9. Gase Systems

9.1. Argon Gas

Argon gas is a noble gas as it is an inert gas. This means that argon does not react with other substances readily and therefore it can provide a useful medium where chemical reactions need to be avoided. Argon creates an inert atmosphere in processes where exposure to oxygen or other reactive gases may lead to a degradation of the chemicals involved. During packaging of pharmaceuticals, argon is sometimes used to displace air and oxygen from the packaging before it is sealed. This helps to extend the shelf life of the pharmaceutical products by reducing the prevalence or rate of oxidative degradation.

The purity of argon is a critical quality attribute as small amounts of impurities or contamination within the gas can jeopardize its strengths and role in preventing chemical reactions and assisting the stability of the product.. Typically, the argon used should be of a high grade, generally 99.999% pure or better, to avoid any contamination. Flow rates and pressure should be considered with particular flow rates and pressure been validated for the intended use.

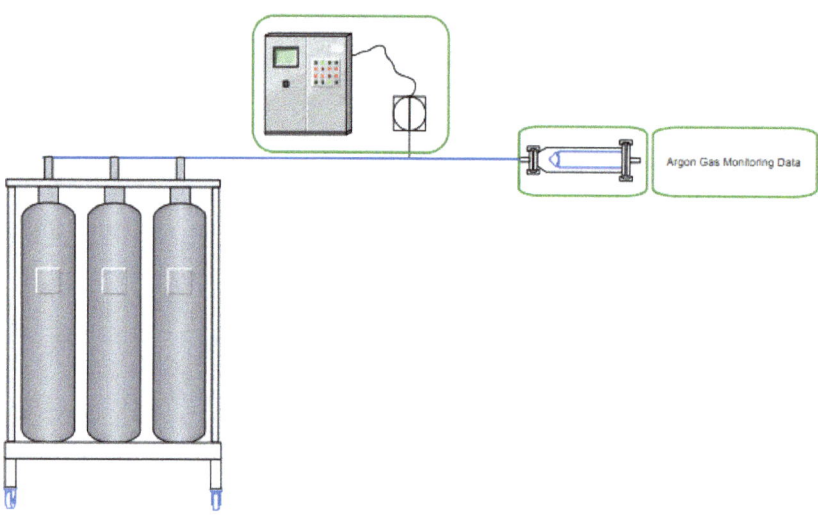

9.2. Carbon Dioxide

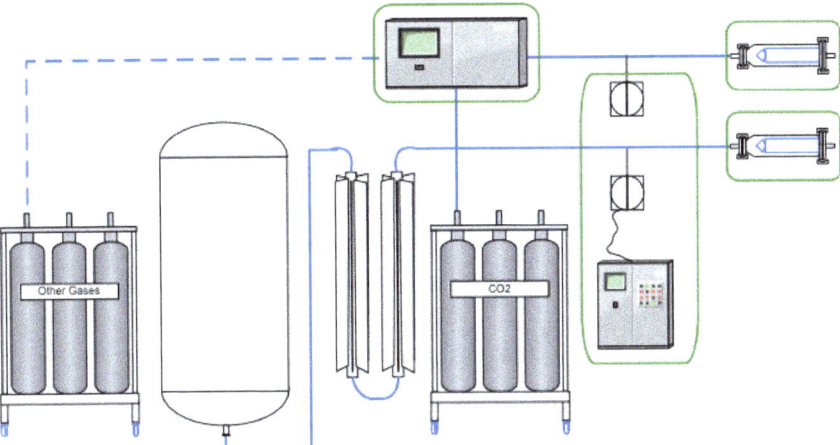

Carbon dioxide (CO2) is used in various capacities in the pharmaceutical and biological sectors due to its unique properties. Here are some prominent applications:

- Purging, inerting/blanketing
- Ph control
- Controlled storage

Purging oxygen from a space, in a similar manner to argon, carbon dioxide can be used to create a relatively inert atmosphere in to prevent oxidation of products. In biological products, CO2 plays an important role in regulating pH levels in cell cultures. Also, for certain biological products, CO2 is used to maintain a controlled atmosphere during storage. This can help in preserving the integrity and efficacy of finished products and help provide a more effective shelf life.

9.3. Oxygen

Equally Oxygen as a gas is widely used in pharmaceutical and biological product manufacturing and production. In contract to inert gases, oxygen can help speed up reactions which may be necessary to achieve a particular outcome or chemical composition. In biological manufacturing, processes such as cell culturing, aeration, fermentation require specific levels of oxygen to support the process and physical mechanisms involved. In some pharmaceutical formulations, oxygen is required to improve stability or other product requirements such as bioavailability.

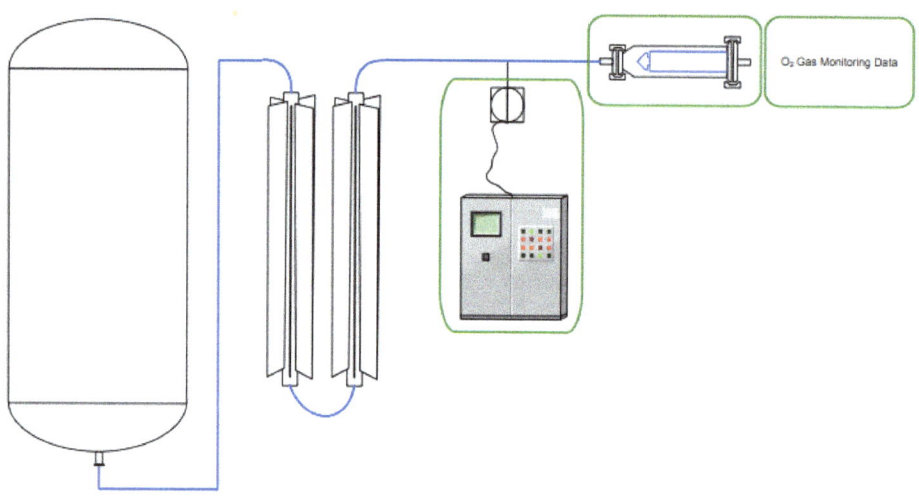

10. Steam Sterilization

Steam Sterilization using moist heat is a long established sterilization method and cost effective. It can provide terminal sterilization for suitable medical devices and their packaging that can be subjected to steam and high temperatures without any impact to the packaging materials or device performance. Sterilization, due to its criticality, requires considerable technical knowledge, planning, testing and verification in order to achieve a sterile product that is safe and effective. This book summarises the verifications and methodologies on offer with regard to steam sterilization.

In addition, some selected factors in relation to Aseptic processing and CGMP is provided for the reader, while not all aspects of aseptic processing are addressed, aseptic processing for some medicinal and healthcare products is an alternative to terminal sterilization. Effective sterilization relies upon the coming together of various disciplines and areas of technical knowledge. Firstly, the design of a Sterilizer needs to comply with regulatory requirements and industry standards. While many steam sterilizers may be off-the-shelf items, their criticality and complexity requires appropriate commissioning, validation, routine monitoring and maintenance.

The steam cycle is monitored by mechanical and biological monitors and the critical parameters are also monitored using a printout of the temperature, dwell times and pressure throughout the cycle. The effectiveness of steam sterilization is monitored with a biological indicator containing spores of Geobacillus stearothermophilus (formerly Bacillus stearothermophilus). Positive spore test results are a relatively rare event and can be attributed to operator error, inadequate steam delivery or equipment malfunction. At normal atmospheric pressure (760 mmHg), boiling temperature is 100°C. However, the temperature must exceed 100°C to kill endospores. The boiling point can be increased by increasing the pressure. By increasing the pressure, the autoclave reaches a boiling point of 100°C or higher (121°C) and kills endospores.

10.1. Sterilization and Disinfection

Within day to day scenarios the words sterilization and disinfection can be seen to be used interchangeably, however, their definitions and real meaning makes them quiet distinct. The accurate definitions and understanding of what they mean is critical in GxP environments and with regard to medical device manufacturing and pharmaceutical operations. The term pathogen in microbiology refers to any microorganisms such as virus', fungi, algae or bacteria. The process of Sterilization is the concerned with the elimination microbes to acceptable levels.

The distinction between sterilization and disinfection can be understood based on the impact the technique has on endospores. Sterilization must kill and completely remove the endospores. In contrast to sterilization, disinfection processes cannot achieve this.

An endospore is a stripped-down, dormant form to which the bacterium can reduce itself and allow bacteria to lie dormant for extended periods, even centuries. They are found in water and soil. Endospores are resistant to ultraviolet radiation, high temperatures and chemical disinfectants, but temperatures above 100°C can destroy them. Antibacterial agents that work by destroying cell walls do not affect endospores. Some classes of bacteria can turn into exospores or microbial cysts. Exospores and endospores are both dormant stages seen in some microorganisms. Some gram-positive bacteria can only make endospores, however, gram-negative bacteria cannot. Sterilization processes, both physical (Steam) or chemical (Ethelene Oxide) destroy pathogens and microorganisms, they are not seen as absolute. Therefore, criteria to determine when medical devices are considered sterile haven been established. To claim effective sterility a reduction ≥106 log in colony forming units (CFU) of the most resistant spores must be achieved from half the proposed regular cycle.

Effective sterilization requires contact of the sterilizing agent with all surfaces of the item to be sterilized. The selection of the agent is based on the nature and type of item to be sterilised, as some medical devices or the packaging materials or design may not allow a particular method of sterilisation. For moist heat sterilization, the critical process parameters include temperature, time and the presence of moisture. Therefore, moisture is the sterilizing agent. Additional parameters such as pressures, temperature changes and dwell times should also be considered in accordance with the applicable ISO standards (ISO 17665-1 for Moist Heat Steam Sterilization) Applying the correct process parameters ensuring minimum sterility is achieved. These can be determined using a parametric approach or a biological approach.

10.2. Parametric Approach

Parametric sterility assurance is when the control of the sterilization process and the 'sterility test' is demonstrated by critical process controls.

> **211.165 (a) Testing and release for distribution.** For each batch of drug product, there shall be appropriate laboratory determination of satisfactory conformance to final specifications for the drug product, including the identity and strength of each active ingredient, prior to release.

> **211.167(a)** For each batch of drug product purporting to be sterile and/or pyrogen-free, there shall be appropriate laboratory testing to determine conformance to such requirements. The test procedures shall be in writing and shall be followed

It applies to terminally sterilized products. For manufacturers, the benefit is that upon meeting the defined sterilization parameters a sterility test does not need to be performed on the product. Also, from a safety and efficacy perspective, it parametric based releasing can provide greater assurance that a batch meets the sterility requirement than can be achieved with a sterility test of finished units (samples) taken from the batch. If Sterility test methodology is applied, as the number of samples taken from a batch tend to be small they may not capture dispersed pathogens. Also, incubation and simulation of growth is time consuming, technical and costly. However, to provide added level of sterilisation verification, A sterilization load monitor, either in the form of a physical, chemical or biological indicator, is included with each load to satisfy the requirement for a laboratory test (sterility test).

Endospore stain of the cell Bacillus subtilis showing endospores as green and the vegetative cell (cells growing) in red

10.3. Risk and Sterility

The impact of contaminated or non-sterile products can result in serious illness or death to patients. For many medical devices, they are placed in the body for extended periods of time or indeed for many years (Hip, Knee impacts, Cardiovascular stents.)In addition injectable sub cutaneous treatments sustain life and bio-chemical processes or treat genetic conditions.

While there is always residual risks associated with the use of medical and medicinal products, there risks are normally acceptable based on very low occurrence or likelihood of occurrence and also as residual risks are assessed in order to be deemed acceptable based on benefit-risk analysis.

Not only is it important to mitigate against any risks throughout the manufacturing process. risk based approach to operations, in particular, changes to the process must be maintained throughout the life cycle of a product. Where appropriate and technically permissible terminal sterilization is the preferred point of sterilization. Terminal sterilization is when the final sealed product in its container sterilised at the end of the process.

- Disinfection and sterilization processes both remove pathogens.
- Sterilization removes pathogens and endospores
- Disinfection is when pathogens are removed but endospores remain
- For Disinfection, it is important to maintain the condition of equipment so that it remains fit for its intended use and is not subject to excessive damage due to the cleaning process.
- Parametric approach to batch release uses documented evidence of the control of critical parameters, hence removing the need to test samples taken from finished product.
- A sterilization load monitor, either in the form of a physical, chemical or biological indicator, is included with each load to satisfy the requirement for a laboratory test (sterility test)

10.4. Spaulding's classification

Spaulding's classification was proposed by the American Physician, Earle H. Spaulding can be used to determine the disinfection or sterilization method that should be chosen according to the medical instrument.

- Low-level disinfection: Instruments that touch intact skin are non-critical items. These require
- High-level disinfection: Instruments that contact incised skin or mucous membranes are semi-critical items. Examples include endoscopes and anaesthesia apparatus. These should undergo high-level disinfection.
- Sterilization: Instruments that touch places where no single microorganism should exist are critical
- items. A representative example is a surgical instrument. These must be sterilized unconditionally.

For the purposes of the manufacture of medical devices that are provided sterile, Spaulding's classification does not apply. The decision to apply disinfection or sterilization applies to hospital settings depending on the equipment and intended use of the equipment. Further to Spaulding's, revision was required to cover the various modern types of treatments and corresponding equipment used that requires anti-microbial techniques e.g. disinfection or sterilization. Such Medical devices that come into contact with the mucous membranes (endoscopes)

The FDA introduced an additional an category of "environmental surfaces" such as walls and floors. Patients do not come into direct contact and therefore these surfaces carry very low-risk of transmission of infection, hence, disinfection frequency can be reduced.

10.5. Cleaning

Cleaning can be defined removing potential contaminants from equipment or product. Cleaning must take place according to defined procedures and programs, with on-going environmental monitoring to ensure compliance to the microbiological limits and to detect the development of resistant strains of organisms.

The effectiveness of all disinfectants must be validated with reference micro-organisms and local isolates. Hard surfaces of equipment, premises and materials that are decontaminated can be selected on a risk based approach. The choice of disinfectants must be adequate to maintain good results on the viable environmental monitoring trend analysis.

Cleaning can be defined as "the process of removing potential contaminants from equipment or product while maintaining the condition of equipment so that it remains fit for its intended use and is not subject to excessive damage due to the cleaning process." The words "grease and dirt free" are useful to remember as they give a practical understanding of what "clean" means. If equipment, parts or product become "dirty" they are often referred to as "soiled". Cleaning is a physical process where, particles, grease or organic matter is removed from a material or surface. NOTE: If equipment is clean it is does not mean it is sterile.

Clean-in-Place often abbreviated to CIP, allows equipment cleaning to occur with minimal disassembly of equipment. CIP programs allow different products using similar or different materials to be manufactured on the same equipment. To comply with Regulatory Requirements, the manufacturer must provide safe and effective products for use. Cleaning compliance is a key part of achieving a state of compliance and more importantly, supplying safe products.

Clean Hold Time (CHT): The total time the parts are held clean post cleaning.

Cleaning Agent: The chemical agent or solution used as an aid in the cleaning process.

Cleaning Process Parameters: The parameters that are critical in the cleaning process. Subsequent cleaning process monitoring may or may not utilise these parameters.

Critical Process Parameter (CPP): A control parameter that has a direct relationship to the quality, safety, effectiveness or performance of the intermediate or final product.

Dirty Hold Time (DHT): The total time the parts (or equipment) are held dirty prior to cleaning.

Maximum Allowable Carry Over (MACO): Amount of allowed product residue carry-over from lot-to-lot and batch-to-batch etc. This limit is based on the lowest of:

Residue: Substance left on surfaces of equipment after cleaning that may pose a risk for subsequent use. Example: residues that may require cleaning include: product, excipients, raw materials/intermediates, non-volatile solvent, non-intrinsic cleaning agents such as detergents etc.

Worst Case Conditions: Considered to pose the greatest chance of process or product failure. The highest or lowest value of a given control parameter or set of parameters.

Visual Inspection: With regards to cleaning, visual inspection should be completed by appropriately trained and experienced personnel on completion of equipment/process clean down. Surfaces should be visibly clean and free of visible residue. Hard to clean places should be examined in particular.

Process Window: The selected operating range of machine setting/parameter that will produce product to meet all quality and product specifications.

Clean-in-Place (CIP): Is a cleaning method used to clean the inner surfaces of piping, vessels and process equipment without the need for disassembly.

PIC/S: The Pharmaceutical Inspection Convention and Pharmaceutical Inspection Co-operation Scheme (referred to as PIC/S) are two international bodies between countries and pharmaceutical inspection authorities, that co-operative in subjects relating to the field of GMP.

Skid: Is essentially a modular process that can be plugged into a process onsite, with little construction or integration. Skids are used as part of Clean-in-Place solutions within the food and beverage and pharmaceutical industries.

10.6. Clean-in-Place (CIP)

Cleaning validation for a CIP system design involves the intersection of two similar or different products. Take a simple example. A pharmaceutical company manufactures two types of paracetamol caplets (tablets). Product A contains the active ingredient paracetamol, preservatives and other excipients. Product B is also a paracetamol product but it contains an additional ingredient, caffeine. Therefore, product B, is branded differently and marketed a more discerning customer in mind. Where multiple products are manufactured on the same equipment or machinery, the process is often referred to as non-dedicated. As with the above example, if the same equipment is used to produce product A and Product B, an intersection of products occurs.

Residue is any substance or trace of substance left on equipment or surfaces after cleaning. It is near possible to remove all residue from surfaces so a residue limit should be medically safe and at a level that does not cause product quality issues or concerns.

Within any cGMP environment, the requirement to maintain a clean and suitable manufacturing area is key to compliance and ensuring product quality and customer safety. Visual inspection of the cleaning process must be done before swabbing. Inspection should confirm the equipment is visually clean and dry and no adverse odours are present. Upon completion of visual inspection, swabbing should then only be carried out if required by procedure. For areas that cannot be accessed for visual inspection or swabbing, a rinse sample can be taken in place of a swab. Sometimes it is not possible to obtain a swab or rinse sample, therefore visual inspection may be the only method used to verify cleaning effectiveness. In any validation an important theme is to challenge the consistency of a process. Samples must be representative to ensure a proper picture is painted. Sampling sites should be taken from "hard to clean" areas as well as "easy to clean" ones to ensure that samples are representative of the equipment.

"Soils" are a source of contamination to products and therefore can present a risk to patients or users. Soils can be introduced by unplanned and unintended events, but they are likely a part of the process or the result of a manufacturing agent being used within a manufacturing process. Examples would include coolant of cutting fluid used in a machining process. With regards to CIP and cleaning between different products cleaning should focus on product contact surfaces or process critical indirect product contact surfaces. Non-critical cleaning of walls, floors and ceilings does not require the same level of cleaning. Likewise, dedicated equipment can often have a reduced cleaning program.

10.6.1. PIC/S Guidance on Limits

The Pharmaceutical Inspection Convention and Pharmaceutical Inspection Co-operation Scheme (jointly referred to as PIC/S) are two international instruments between countries and pharmaceutical inspection authorities, which provide together an active and constructive co-operation in the field of GMP.

I. No more than 0.1% of the normal therapeutic dose of any product should appear in the maximum daily dose of the following (next) product.

II. No more than 10 (parts per million, ppm) of any product will appear in another product, (this value is not always the default).

III. No quantity of residue should be visible on the equipment after cleaning procedures are completed. Spiking studies should determine the concentration at which most active ingredients are visible.

10.7. Antimicrobial Techniques

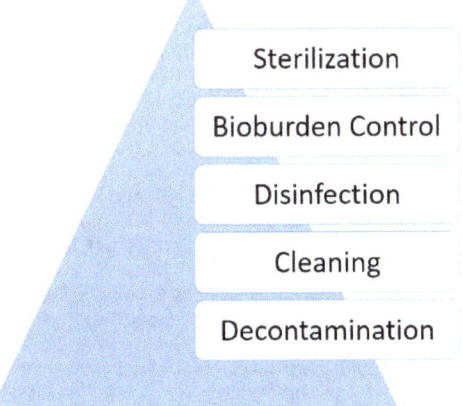

10.7.1. Pasteurization

The process of heating product in properly designed and operated equipment a specified temperature and held continuously at or above that temperature for a specified time is known as Pasteurization. It is primarily in the food and beverage industry such as Milk pasteurization which intends to reduce transmission of communicable diseases and reduce harmful pathogens.

Pasteurization	Sterilization
• Partial elimination of pathogens • Temperatures below 100°C	• Complete elimination of pathogens (according to SAL) • Temperatures in excess of 100°C

Pasteurization is not an end point for medical device products as it does not meet the safety requirements or requirements for sterile use. However, during manufacturing or process of raw materials, pasteurization steps may be applied to control bioburden and microbial levels.

10.8. Sterilisation Processes

Sterilization processes must demonstrate that they deliver products that are free of viable microorganisms, thus making them safe and effective. The initial development efforts in selecting an appropriate technology and sterilization technique is a critical part of product design and development. Once a product is designed (e.g. medical device-Orthopedic implant) and successful developed and transferred into the manufacturing space, critical to quality and safety processes such as sterilization must be maintained during routine use. Healthcare Authorities apply a number of processes and methodologies that work to ensure medical devices and healthcare products are safe and effective. Likewise, sterilization processes fall under their remit as they regulate products and the markets which they have legislative responsibility. (e.g. MHRA in the UK is the component authority, FDA in the United States.

For Sterilization Processes the manufacturer of medical devices must meet several requirements and is responsible for the safety of products on the market. Healthcare authorities seek evidence and technical documentation that contributes to a manufacturer been granted market approval. These include, but are not limited to:

- Manufacturer of medical devices or healthcare products are subject to license by regulatory bodies
- Organizations (e.g. manufacturer) are subject to audit by regulatory bodies and notified bodies (e.g. BSI)
- Regulatory processes requiring registration and approval of products
- Organizations must demonstrate compliance a Quality Management system or Quality System.
- Qualification and Validation of facilities, equipment and ancillary equipment
- Manufacturers must comply with International Standards (e.g. CEN, ISO-17665 1-3)
- Regulatory processes requiring registration and approval of products
- Organizations must demonstrate compliance a Quality Management system or Quality System.
- Qualification and Validation of facilities, equipment and ancillary equipment
- Manufacturers must comply with International Standards (e.g. CEN, ISO-17665 1-3)
- ISO 17665-1, Sterilization of health care products - Moist heat - Part 1: Requirements for the development, validation and routine control of a sterilization process for medical devices
- ISO-17665-2 Sterilization of health care products - Moist heat - Part 2: Guidance on the application of ISO 17665
- ISO-17665-3 Sterilization of health care products - Moist heat - Part 3: Guidance on the designation of a medical device to a product family and processing category for steam sterilization

10.9. FDA Categorisation of Established Sterilization Processes

Two categories of sterilization methods recognised by the FDA for sterilize medical devices in manufacturing settings are (1) established and (2) novel.

Established Sterilization Methods:

1. Established Category A: Established Category A Definition

> These are methods that have a long history of safe and effective use as demonstrated through multiple sources of information such as ample literature, clearances of 510(k)s or approvals of premarket approval (PMA) applications, and satisfactory QS inspections. For established methods such as dry heat, EO, steam, and radiation, there are voluntary consensus standards for development, validation, and routine control that are recognized by FDA.

Established Category A Sterilization Methods:

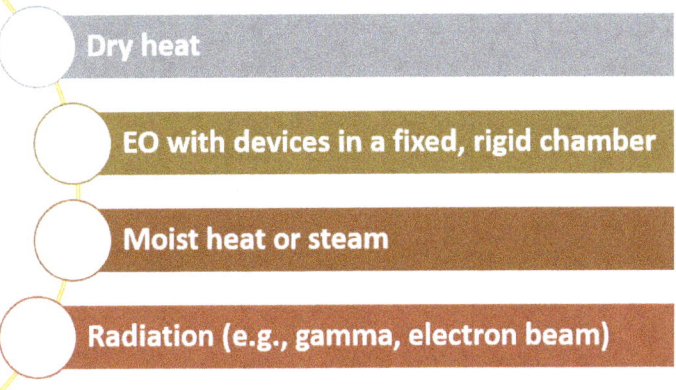

- Dry heat
- EO with devices in a fixed, rigid chamber
- Moist heat or steam
- Radiation (e.g., gamma, electron beam)

Established Category B Definition

> There are other established methods for which there are no FDA-recognized dedicated consensus standards, but for which published information on development, validation, and routine control is available. In cases where FDA has previously evaluated sterilization development and validation data for specific sterilizers using discrete cycle parameters and determined the validation methods to be adequate, we consider these to be Established category B

Examples of these Established Category B Sterilization Methods:

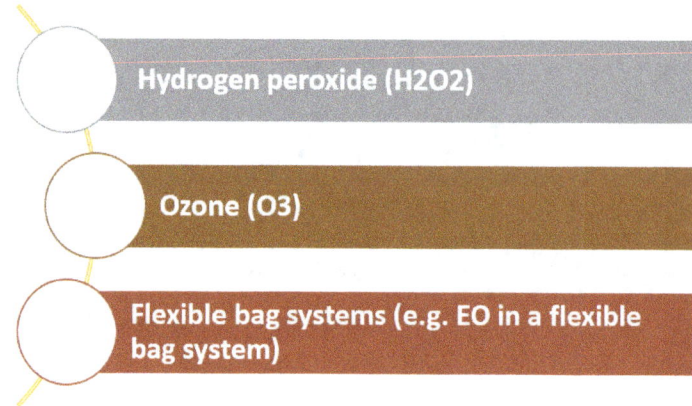

An additional category under Novel methods accounts for methods where the specific process does not appear to have been evaluated by FDA or because process validation data has not been evaluated and found to be adequate in previous cleared or approved submissions, and are therefore considered novel.

Note: Novel Sterilization Methods are newly developed methods for which there exists little or no published information, no history of comprehensive FDA evaluation of sterilization development and validation data through an FDA-cleared 510(k) or approved PMA for devices sterilized with such methods, and no FDA-recognized dedicated consensus standards on development, validation, and routine control. A Novel Sterilization Method is a method that FDA has not reviewed and determined to be adequate to effectively sterilize the device for its intended use.

FDA also considers methods where the specific process does not appear to have been evaluated by FDA, either because the parameters of an FDA-cleared sterilizer have been altered, or because process validation data have not been evaluated and found to be adequate in previous cleared or approved submissions, to be novel. Examples of Novel Sterilization Methods include Vaporized peracetic acid, High intensity light or pulse light, Microwave radiation, Sound waves, and Ultraviolet light.

10.10. Steam Sterilizer (Moist Heat) - Development of Sterilization Processes

Terminal Sterilization is a well-established method of sterilizing medical devices in their primary packaging configuration. Blister packaging is often designed to withstand the temperatures and conditions presented under steam sterilization via moist heat. In many companies, sterilization processes can be long established and in use for commercial products been released to the market. Therefore, the requirement to develop sterilization processes arises where a new product type or new product offering is planned. According to the regulatory requirements, medical devices must be developed in a structed manner that is often referred to as product realization or product development. Specific requirements are detailed in ISO 13485 and has specific requirements under the Product realization clause. 21 CFR Part 820.30 Design controls sets out the requirements of development planning, design inputs, design outputs, design verification and validation. The process of sterilization is subject to these controls and requirements. Factors to consider in development of Sterilization (Moist Heat) Process:

- Is there a Quality Management system established?
- Is the product material suitable for moist heat sterilization?
- Are there potential contaminants in the sterilization agent (moist heat)?
- Is the product a solid, liquid or combination of both (e.g. contact lens system)?
- What is the packaging materials to be used (blister materials, container, Lid stock)?
- Is the nominal mass or volume of the product known?
- Is the product manufactured with existing equipment and validated processes?
- Are the sterilization conditions (e.g. temperature cycle time) known?
- Shall the product be subject to multiple sterilization cycles?

Note: The microbial effectiveness of moist heat used in the process of steam sterilization is well established and documented in published literature and scientific journals.

10.10.1. Instrument/Equipment Design

The design and shape of instruments that requires sterilization can influence the steam and heat penetration. Does it have pin joints? (e.g. scissors). Does it have moving parts? (e.g. fasteners) and so on. Porous materials make it difficult to remove air during the sterilization process. For forceps or scissors, they may need to be the open position to ensure effective exposure during sterilization.

10.10.2. Material

The materials of construction of instruments or sterilizable equipment needs to be compatible with the intended method of sterilization. In addition, any processes that seal, bond, assemble materials into the product configuration needs to be assessed in relation to their suitability for the sterilization method used. The properties of materials and their thermal conductivity also influences how hard they are to "heat up" or allow heat transfer (moist steam sterilization)

10.10.3. Product Family

A product family is a method of grouping product based on certain attributes. With the attributes selected the steam penetration resistance can be estimated.

10.10.4. Processing Parameters

The critical variables for Sterilization (moist heat) is temperate, time and moisture presense. For parametric testing, the below tests can be applied to Sterilizers.

- Hollow load test
- Standard pack test
- Thermometric test
- Bowie and Dick test
- Air leakage flow rate test
- Air detector test
- Load dryness
- Sounds power test
- Dynamic pressure test
- Steam quality tests
- Water
- Compressed Air

The data harvasted from these tests can be assessed to determined if the sterilization process has been achieved.

While the critical process variables are time, pressure and moisture. A sterilization cycle has different stages in which these variables need to meet such as target temperatures and pressures at which they then need to be maintained for a period od time (dwell time). Each of these temperatures and pressures have tolereances also. As absolute accuracy is not achieveable. The data from the tests listed above are used to established the process parameters

10.11. The Sterilizer as Equipment

While the object of sterilization equipment is to render medical devices safe and sterile, there is a responbility to ensure they operate safely. The impact on the envirnment is mostly governed by national enviromental legislation and should also be considered by manufacturers using steam sterilizers. Steam sterilization uses water in both its liquid and vapourrs state to penetrate as steam into a load (term used for medical devices when insitu in the sterilizer) at which point it condenses onto the surfaces of the device or the primary packaging of the device. The sterilization process needs to operate consistently in order to provide manufacturers and patients with the protect required. This requirement of consistency or reproduceability can be impacted by:

   I. Change in specified cycle parameters
   II. Orientation of the load
   III. Overheating within the system
   IV. Excessive build up of condensate

The quantity of data points or signal inputs is an important consideration for the temperature and pressure measuremen during validation activities and if routine testing is required. Thermocouples can be positioned throughout a chamber. The use of KAYE validators can facilitate many signal inputs (thermocouples) with some configurations allowing 36 independent points of measurement.

Hollow load test

The test uses a hollow test piece per EN 285. It is essentially a test for steam penetration into the medical device. The standard pack test result is assessed based on the exposure to a chemical indicated that is inserted in the load test piece.

Standard test pack

The standard test pack is used for:

   I. Small load test
   II. Full load test
   III. Bowie and Dick test
   IV. Air detector test
   V. Load dryness test (textiles)

A standard test pack is made up of plain cotton sheets of a specific size, number of threads per centimetre and weight.

Thermometric tests

The small load thermometric test is a test for steam penetration into a standard test pack. A number of thermocouples or temperature sensors are located a different levels within the standard test pack over the vertical axis.

Bowie-Dick Test (Steam Penetration)

Steam penetration is tested using a Bowie Dick test kit. To verify the consistency of the process, this is typically done three times for a recipe or cycle. It is a similar test to the small load test. Bowie dick test should be completed with reference to ISO 11140-3, Sterilization of health care products — Chemical indicators — Part 3: Class 2 indicator systems for use in the Bowie and Dick-type steam penetration test

Steam penetration is tested using a Bowie Dick test kit. To verify the consistency of the process, this is typically done three times for a recipe or cycle. The Bowie-Dick test is used to detect air leaks and inadequate air removal and consists of folded 100% cotton surgical towels that are clean and preconditioned. A commercially available Bowie-Dick-type test sheet should be placed in the center of the pack. The test pack should be placed horizontally in the front, bottom section of the sterilizer rack, near the door and over the drain, in an otherwise empty chamber and run at 134°C for 3.5 minutes. The test is used each day the vacuum-type steam sterilizer is used, before the first processed load. Air that is not removed from the chamber will interfere with steam contact. Smaller disposable test packs (or process challenge devices) have been devised to replace the stack of folded surgical towels for testing the efficacy of the vacuum system in a prevacuum sterilizer. These devices are "designed to simulate product to be sterilized and to constitute a defined challenge to the sterilization process."

They should be representative of the load and simulate the greatest challenge to the load. Sterilizer vacuum performance is acceptable if the sheet inside the test pack shows a uniform color change. Entrapped air will cause a spot to appear on the test sheet, due to the inability of the steam to reach the chemical indicator. If the sterilizer fails the Bowie-Dick test, do not use the sterilizer until it is inspected by the sterilizer maintenance personnel and passes the Bowie-Dick

Air leakage test
The purpose of air leakage testing is to verify that the chamber is vacuum-tight and can maintain the vacuum over a period of time. To avoid loose interpretations, a formal definition of vacuum-tight should be documented. The British standard EN 285+A2 "Sterilisation. Steam sterilisers. Large sterilisers" provides definitions, guidance and a framework for testing steam sterilisers. The air leakage test should result in the chamber maintaining a predetermined pressure over a set period of time e.g. ten minutes.

Load Dryness
The purpose of load dryness tests is verify the level of moisture remaining in the load at the end of sterilization cycle is acceptable. E.g. 0.2 % for metals.

Sound power test
This refers to the sound power generated from the sterilizer. It is a environmental style test.

10.12. The Sterilization Process

Autoclaves or steam sterilisers are used to sterilise items such as tools, fixtures and utensils used in aseptic processing. Modern systems are designed to fulfil the requirements of FDA and EU regulatory requirements. DIN 58950/58951 is a standard in which many manufacturers design and build steam sterilisers to fulfil the requirements set out in the document. Conformance to this standard ensures autocalves comply with the FDA and GMP directives. Industrial steam steriliser systems used in biotechnology companies comprise the following main components:

   I.   Pressure container for sterilisation
   II. Vacuum pump
   III. PLC controller
   IV. Human Machine Interface (HMI)
   V.   Cycle software

The sterilisation process can be divided into three distinct stages:

- Pre-treatment Stage: during this stage the autoclave begins to heat up and the air in the chamber is replaced by a mixture of steam and air.
- Sterilisation Stage: the purpose of this stage is to kill any harmful microbes by using steam sterilisation. The temperature and pressure of the chamber is held at predefined settings for a specific period of time.
- After-treatment Stage: cooling, decompression and drying occurs in this stage of the cycle.

The steriliser can be loaded with the help of a loading trolley manufactured with suitable materials or an automatic loading and unloading system. The steriliser can alternatively be equipped with trays for accommodating the goods to be sterilised.

Pressure Leak Testing
This test is used to ensure the chamber does not leak. During the course of the test, the pressure is trended. The pressure drop over the test must be within specification. For example, the pressure decrease should be less than 100mbar during the course of the cycle (e.g. 10 minutes).

Mechanical Indicators
Sterilizers have gauges, thermometers, timers, recorders, and/or other devices that monitor their functions. Most sterilizers have automatic controls and alarm systems that are activated if the sterilizer fails to operate correctly.

Chemical Indicators
A chemical indicator on a package verifies exposure to a sterilization process. This helps differentiate sterilized from unsterilized items. It also helps monitor physical conditions within the sterilizer to alert personnel if the process has been inadequate. An indicator may be placed inside a package in a position most likely to be difficult for the sterilant to penetrate. A chemical indicator can detect sterilizer malfunction or human error in packaging or loading the sterilizer. If a chemical reaction on the indicator does not show expected results, the item should not be used.

Biological Indicators
Biological indicators (BIs) are test units that contain viable microorganisms that allow a known microbial challenge to be subjected to a specified sterilization process. Some key Industry standard and subparts that provide clear requirements for the proper use of BIs include:

ISO 11138- Sterilization of health care products,

> Part 1: General requirements
> Part 2: Biological indicators for ethylene oxide sterilization processes
> Part 3: Biological indicators for moist heat steam sterilization processes
> Part 4: Biological indicators for dry heat sterilization processes
> Part 5: Biological indicators for low temperature steam and formaldehyde sterilization processes

Term	Definition
Sterility Assurance Level (SAL)	The probability of a viable microorganism being present on a product unit after sterilization, normally expressed as 10^{-n}
D-value	Time or dose required to achieve inactivation of 90% of a population of test microorganisms under dose conditions
F_{BIO} value	The product of the logarithm of the population and the D-value where F_{BIO} value is an expression of the resistance of the biological indicator
Inoculated carrier	Supporting material on or in which a defined number of viable test organisms have been deposited
Rapid Readout Biological Indicator	Biological indicator with means (typically Enzymatic) for early detection of sterilization failures with a high degree of correlation to the growth of the test organisms

The load monitor(s) should be placed in appropriate positions to indicate that the load was exposed to a sterilization process which was measured and recorded for conformance with defined criteria for parametric release. This position(s) is determined based on the evaluation of development and qualification data. The location and number of monitors should be described and justified in the application. Alternative procedures for demonstrating that a load or part of a load was exposed to a sterilization process should be discussed with the review division(s) prior to submitting a plan for parametric release.

ISO 11138: Sterilisation of healthcare products—biological indicators
Positive assurance that sterilization conditions have been achieved can be obtained only through a biologic control test. The biologic indicator detects non sterilizing conditions in the sterilizer. A biologic indicator is a preparation of living spores resistant to the sterilizing agent. These may be supplied in a self-contained system, in dry spore strips or discs in envelopes, or sealed vials or ampoules of spores to be sterilized and a control that is not sterilized. Some incorporate a chemical indicator also. A biologic indicator must conform with USP testing standards. A control test must be performed at least weekly in each sterilizer. Many hospitals monitor on a daily basis; others test each cycle.

10.13. Validation of Steam Sterilizers

10.13.1. Installation Qualification (IQ)

Installation Qualification (IQ) is establishing by objective evidence that all key aspects of the process equipment and ancillary system installation adhere to the manufacturer's approved specification and that the recommendations of the equipment supplier have been suitably considered.

IQ is required for new equipment. IQ is an element of the validation lifecycle which must be completed prior to the Operational Qualification (OQ) of equipment or a system. In terms of verifying the equipment is manufactured, assembled and operates to its design requirements and intended use, factory acceptance testing (FAT), commissioning and Site Acceptance Testing (SAT) is regarded as pre-requisite activities before equipment validation, in particular for critical to quality systems large equipment and complex/automated conditions. If changes are required to an existing and qualified sterilizer, then requalification (IQ/OQ) may be required e.g. change or replacement of door seals, pump replacements and so on.

After the design and specification and construction of equipment is completed, it is the function of Equipment validation to verify all requirements are fulfilled and provide documented evidence of the verification activity. Installation Qualification verifies fundamental requirements of equipment are met. Vendor quotations, vendor manuals and documentation, user requirements specifications, design specifications, regulatory requirements, International standards can all be inputs into developing an Equipment validation strategy that includes Installation Qualification (IQ) verifications and Operational Qualifications and so on.

- Documented version of the Sterilization cycle operating parameters and settings as programmed into the control system
- Operating instructions
- Instruction on if a process parameter fails to reach its requirement and how this can be identified and a method of identifying an error in the control and how this can be identified.
    - E.g. a printed paper record of the cycle can be completed to ensure the requirements of the cycle have been fulfilled when reviewed against a pre-approved specification
    - E.g. Engineering controls
    - E.g Quality review
- BI verification
- Calibration instructions
- Maintenance requirements

10.13.2. Operational Qualification (OQ)

Operational qualification (OQ) is defined as providing documented evidence that the equipment operates consistently across its full operating range and meets the specifications detailed in the user requirement specification. For Steam Sterilizers, the equipment must be capable of operating to the requirements and parameters specified for the sterilisation cycle. The OQ also ensure that's the entire chamber is assessed for the distributed parameters such as a temperature to verify there are no cold spots in the chamber of autoclave. Temperature distribution therefore involves multiple thermocouples or temperature sensors across the whole chamber and is carried out empty. OQ should cover the following providing evidence of successful verification:

- safety features and mechanisms function as design and intended
- the equipment operates according to pre-determined limits
- the operating cycle (sterilization cycle or program is delivered as intended
- no interference from other equipment during the cycle
- sound pressure at the installation site meets legal requirements

In order to provide confidence in the equipment and demonstrate consistency, a number of OQ cycles can be completed:

I. OQ- Minimum load in chamber
   a. Cycle 1
   b. Cycle 2
   c. Cycle 3
II. OQ- Maximum load in chamber
   a. Cycle 1
   b. Cycle 2
   c. Cycle 3

For Steam Sterilizers, the effectiveness of sterilization can be impacted by air leakage into the chamber. OQ testing should also include the following tests for parametric approach:

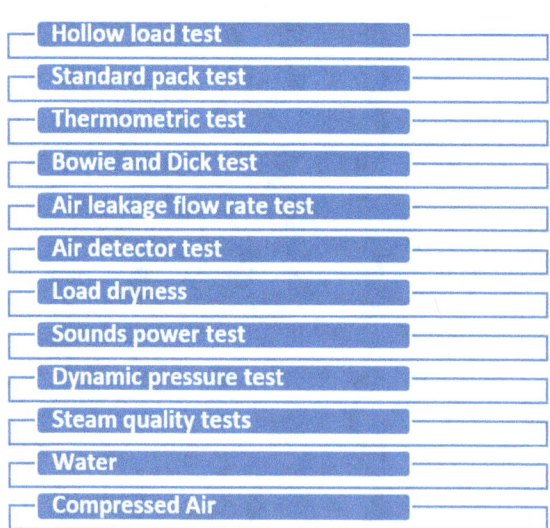

10.13.3. Performance Qualification (PQ)

In general validation terms, Performance Qualification (PQ) is defined as establishing confidence through appropriate testing that finished product produced by specified process meet all release requirement and pre-defined specifications. Performance Qualification for Steam Sterilizers confirms that the product has been exposed to the sterilization process using the equipment. From a Steam Sterilization perspective PQ is designed to demonstrate that the sterilization process is capable of achieving a specified sterility assurance level (SAL) for a load on a consistent and repeatable basis.

The PQ should:

I. Ensure IQ/OQ completed successfully prior to PQ

II. the standard/routine sterilisation cycle

III. packaging should reflect the standard/routine packaging

IV. the load configuration used for production

V. The number and location of temperture probes/sensors should be documented with rationale to demonstrate effective sterilization throughout the chamber.

VI. Biological indicators, their quantity and position should be documented

VII. Demonstrate a minumum of 3 consecutive sterilisation cycles that meet all requirements

VIII. PQ meets the requirements of regulation, standards required to suupport

10.14. Requalification

The necessity for requalification is (1) mandated according to regulatory requirements and state of the art approaches detailed industry Sterilization standards (e.g. annual requalification) (2) changes that may impact the performance or safety of the product:

-Change in the packaging materials type
-Change in the manufacturing process- equipment or operating conditions (e.g. packaging of the product)
-Consider re-qualification when significant changes are made to significant change in the product material or design tolerances that could affect the heating rate of the material
-Change in the sterile barrier venting
-Changes that could affect microbial barrier efficacy,
-Changes in packaging design,
-Changes in vendors that could significantly affect physical properties and heat transfer
-Changes that could affect ability to maintain specified operating parameters or that could substantially change the rate of heat transfer or steam penetration into the product
-Process changes that could substantially change the manner in which process parameters are reached, controlled or controlled (e.g. such as new software or new hardware)
-Change in product loading or density/ changes in the previously validated loading configurations
-Under appropriate circumstances it may be possible to document equivalency between sterilizers to allow the processing of product in a given sterilizer based on documented equivalency to another sterilizer where the product or designated master product from the same product family was previously validated.

10.15. Industry Standards relevant to Sterilization

Industry Standard
ANSI/AAMI St 67: Sterilization of health care products-Requirements and guidance for selecting a sterility assurance level (SAL) for products labeled as "sterile"
ISO 18472: Sterilization of health care products –Biological and chemical indicators – Test Equipment

Industry Standard
EN 556-1: Sterilization of Medical Devices - Requirements for Medical Devices to Be Designated "Sterile" - Part 1: Requirements for Terminally Sterilized Medical Devices

Schematic of Steam Sterilizer

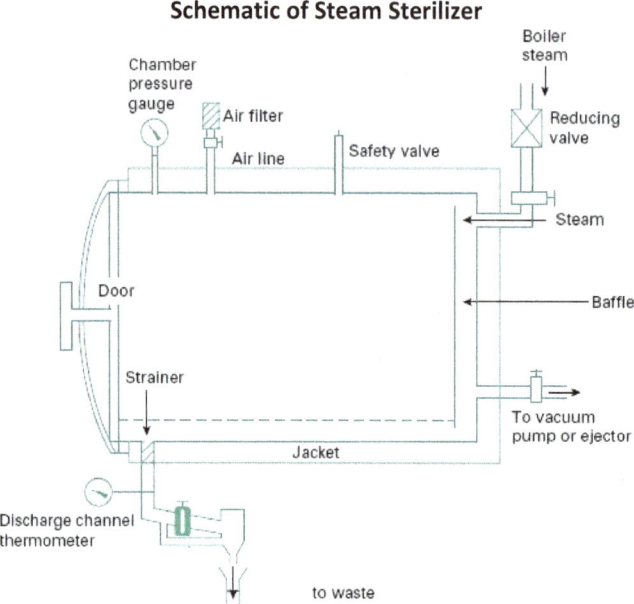

Mechanical Indicators
Sterilizers have gauges, thermometers, timers, recorders, and/or other devices that monitor their functions. Most sterilizers have automatic controls and alarm systems that are activated if the sterilizer fails to operate correctly.

Chemical Indicators
A chemical indicator on a package verifies exposure to a sterilization process. This helps differentiate sterilized from unsterilized items. It also helps monitor physical conditions within the sterilizer to alert personnel if the process has been inadequate. An indicator may be placed inside a package in a position most likely to be difficult for the sterilant to penetrate. A chemical indicator can detect sterilizer malfunction or human error in packaging or loading the sterilizer. If a chemical reaction on the indicator does not show expected results, the item should not be used.

Biological Indicators
Biological indicators (BIs) are test units that contain viable microorganisms that allow a known microbial challenge to be subjected to a specified sterilization process. Some key Industry standard and subparts that provide clear requirements for the proper use of Bis include:

ISO 11138- Sterilization of health care products,
Part 1: General requirements
Part 2: Biological indicators for ethylene oxide sterilization processes
Part 3: Biological indicators for moist heat steam sterilization processes
Part 4: Biological indicators for dry heat sterilization processes
Part 5: Biological indicators for low temperature steam and formaldehyde sterilization processes

Term	Definition
Sterility Assurance Level (SAL)	The probability of a viable microorganism being present on a product unit after sterilization, normally expressed as 10^{-n}
D-value	Time or dose required to achieve inactivation of 90% of a population of test microorganisms under dose conditions
F_{BIO} value	The product of the logarithm of the population and the D-value where F_{BIO} value is an expression of the resistance of the biological indicator
Inoculated carrier	Supporting material on or in which a defined number of viable test organisms have been deposited
Rapid Readout Biological Indicator	Biological indicator with means (typically Enzymatic) for early detection of sterilization failures with a high degree of correlation to the growth of the test organisms

The load monitor(s) should be placed in appropriate positions to indicate that the load was exposed to a sterilization process which was measured and recorded for conformance with defined criteria for parametric release.

This position(s) is determined based on the evaluation of development and qualification data. The location and number of monitors should be described and justified in the application. Alternative procedures for demonstrating that a load or part of a load was exposed to a sterilization process should be discussed with the review division(s) prior to submitting a plan for parametric release.

ISO 11138: Sterilisation of healthcare products—biological indicators

Positive assurance that sterilization conditions have been achieved can be obtained only through a biologic control test. The biologic indicator detects nonsterilizing conditions in the sterilizer.

A biologic indicator is a preparation of living spores resistant to the sterilizing agent. These may be supplied in a self-contained system, in dry spore strips or discs in envelopes, or sealed vials or ampoules of spores to be sterilized and a control that is not sterilized. Some incorporate a chemical indicator also.

A biologic indicator must conform with USP testing standards. A control test must be performed at least weekly in each sterilizer. Many hospitals monitor on a daily basis; others test each cycle.

Very load of implantable devices must be monitored and the implant should not be used until negative test results are known.

10.16. Principle of Operation

Autoclaves or steam sterilisers are used to sterilise items such as tools, fixtures and utensils used in aseptic processing. Modern systems are designed to fulfil the requirements of FDA and EU regulatory requirements. DIN 58950/58951 is a standard in which many manufacturers design and build steam sterilisers to fulfil the requirements set out in the document. Conformance to this standard ensures autoclaves comply with the FDA and GMP directives. Industrial steam steriliser systems used in biotechnology companies comprise the following main components:

- Pressure container for sterilisation
- Vacuum pump
- PLC controller
- Human Machine Interface (HMI)
- Cycle software

The sterilisation process can be divided into three distinct stages:

Pre-treatment Stage: during this stage the autoclave begins to heat up and the air in the chamber is replaced by a mixture of steam and air.

Sterilisation Stage: the purpose of this stage is to kill any harmful microbes by using steam sterilisation. The temperature and pressure of the chamber is held at predefined settings for a specific period of time.
After-treatment Stage: cooling, decompression and drying occurs in this stage of the cycle.

The steriliser can be loaded with the help of a loading trolley manufactured with suitable materials or an automatic loading and unloading system. The steriliser can alternatively be equipped with trays for accommodating the goods to be sterilised.

Air Leakage Test (Vacuum) Test
The purpose of air leakage testing is to verify that the chamber is vacuum-tight and can maintain the vacuum over a period of time. To avoid loose interpretations, a formal definition of vacuum-tight should be documented. The standard EN285 "Sterilisation. Steam sterilisers. Large sterilisers" provides definitions, guidance and a framework for testing steam sterilisers. The air leakage test should result in the chamber maintaining a predetermined pressure over a set period of time e.g. ten minutes.

Steam Penetration (Bowie Dick) Test
Steam penetration is tested using a Bowie Dick test kit. To verify the consistency of the process, this is typically done three times for a recipe or cycle.

The Bowie-Dick test is used to detect air leaks and inadequate air removal and consists of folded 100% cotton surgical towels that are clean and preconditioned. A commercially available Bowie-Dick-type test sheet should be placed in the center of the pack.

The test pack should be placed horizontally in the front, bottom section of the sterilizer rack, near the door and over the drain, in an otherwise empty chamber and run at 134°C for 3.5 minutes. The test is used each day the vacuum-type steam sterilizer is used, before the first processed load. Air that is not removed from the chamber will interfere with steam contact. Smaller disposable test packs (or process challenge devices) have been devised to replace the stack of folded surgical towels for testing the efficacy of the vacuum system in a prevacuum sterilizer. These devices are "designed to simulate product to be sterilized and to constitute a defined challenge to the sterilization process."

They should be representative of the load and simulate the greatest challenge to the load.835 Sterilizer vacuum performance is acceptable if the sheet inside the test pack shows a uniform color change. Entrapped air will cause a spot to appear on the test sheet, due to the inability of the steam to reach the chemical indicator. If the sterilizer fails the Bowie-Dick test, do not use the sterilizer until it is inspected by the sterilizer maintenance personnel and passes the Bowie-Dick test.

Pressure leak testing
This test is used to ensure the chamber does not leak. During the course of the test, the pressure is trended. The pressure drop over the test must be within specification. For example, the pressure decrease should be less than 100mbar during the course of the cycle (e.g. 10 minutes).

11. Alternatives to Steam Sterilization

11.1. Ethylene Oxide (EO)

Chemical sterilants provide an alternative method for sterilizing heat sensitive items if steam sterilization of moist heat can impact the safety or performance of the device. If steam sterilization is not preferred, validation of the sterilization process is still necessary and subject to the relevant regulatory requirements and standards for the method of sterilization. EO gas sterilization is dependent upon four parameters:

    I.   EO gas concentration
    II.   Temperature
    III. Humidity
    IV. Exposure time

ETO is a colourless gas that is flammable and explosive. Each parameter may be varied. Consequently, EO sterilization is a complex multi-parameter process. Each parameter affects the other dependent parameters. The four essential parameters ranges are:

- gas concentration (450-1200 mg/l)
- temperature (37-63°C)
- relative humidity (40-80%)
- exposure time (1 to 6 hours)

Ethylene oxide is a cyclic ether and is used for sterilization of critical items such as plastics, which cannot withstand high temperatures. Ethylene oxide penetrates well into the cell, reaching the DNA of the microorganism and killing it by alkylation. However, ethylene oxide can be harmful to the human body, and sterilization time can take several hours.

Chemical sterilant: These are chemicals used for a longer duration (3–12 h) to destroy all forms of microbes, e.g., peracetic acid (PAA) (0.2%), and hydrogen peroxide (7.5%).

Disinfection: Disinfection is defined as a process of complete elimination of vegetative forms of microorganisms except the bacterial spores from inanimate objects. Technically, there is reduction of ≥103 log CFU of microorganisms by this method without spores.

The use of ETO evolved when few alternatives existed for sterilizing heat- and moisture-sensitive medical devices; however, favorable properties account for its continued widespread use. Two ETO gas mixtures are available to replace ETO-chlorofluorocarbon (CFC) mixtures for large capacity, tank-supplied sterilizers. The ETO-carbon dioxide (CO_2) mixture consists of 8.5% ETO and 91.5% CO2. The main disadvantages associated with ETO are the lengthy cycle time, cost, and its potential hazards to patients and staff; the main advantage is that it can sterilize heat- or moisture-sensitive medical devices. The basic ETO sterilization cycle consists of five stages:

- preconditioning and humidification,
- gas introduction,
- exposure,
- evacuation,
- air washes

As ETO is absorbed by materials, the items or load must undergo aeration to remove residual ETO.

11.2. Oxidizing and Non Oxidizing Disinfectants

Disinfectants are typically categorized into oxidizing agents and non-oxidizing agents. The key distinction is in how they 'work' or the mechanism in which they rendor or destroy pathogens. Oxidising agents destroy micro-organisms while non oxidising disinfectants act as coagulators. Examples of oxidizing agents include peroxide, sodium hypochlorite and iodine. Non-oxidizing agents or coagulating agents include alcohol and ethylene oxide. A summary of the mechanism of non-oxidizing agents entails reacting with microorganisms and cross-linking all the ingredients to coagulate.

11.3. Sodium hypochlorite

Sodium hypochlorite is an commonly used oxidant. At concentrations high enough, over 41% (> 500 ppm), it can corrode metals. Sodium hypochlorite should be carefully handled due to its irritation to the mucous membranes, and thorough rinsing should be performed after use. It must be mixed with water for use. Use of hypochlorite with disinfectants other than water can be dangerous and presents health and safety challenges. Sodium hypochlorite does not kill spores at routine concentrations, but it has a sporicidal effect at 5,000 ppm or higher concentrations.

12. Depyrogenation

Depyrogenation is a thermal process that involves the removal of pyrogens from components (e.g. vials or containers) that are used for injectable pharmaceuticals and biopharmaceuticals. A pyrogen is defined as any substance that can cause a fever. Bacterial pyrogens include endotoxins. Later on we shall see that endotoxins are used to challenge depyrogenation tunnels. Depyrogenation tunnel design varies depending on the manufacturer, however, they usually consist of the following components:

   I.   Infeed and preheating
   II. Heating zone
   III. Cooling zone
   IV. Outfeed and transport to next unit operation
   V.   Automatic emptying

12.1. Pyrogens

Pyrogens are fever inducing proteins of low-molecular-weight proteins. Pyrogens of external origin are referred to as exogenous pyrogens. Modern injection and delivery systems are largely safe, yet adverse reactions are still reported. If a treatment or medication administered via hypodermic needle is contaminated with toxins such as pyrogens fever can be induced which can lead to some death in some cases. It was known in the latter part of the 19th century that some parenteral solutions caused a marked rise in body temperature.

The fever producing agents were not known, and hence described in general terms such as "injection fever," "distilled water fever," and "saline fever". Bacterial pyrogens are responsible for many of those early fevers and for many of the other biological effects described incidental to parenteral therapies. The route of administration of a drug allows a pyrogen, if present, to bypass the bodies primary defences.

The host's response is mediated through the leukocytes (white blood corpuscles) which in turn release their own kind of pyrogen (endogenous pyrogen) and this in turn initiates a fever like response and other biological reactions.

12.2. Bacterial Toxins

There are two general kinds of bacterial toxins: (1) endotoxins and (2) exotoxins. Endotoxins can be extracted from a wide variety of gram-negative bacteria. The term "endotoxin" is usually interchangeable with the term "pyrogen" although not all pyrogens. Higher doses of endotoxin are required to produce a lethal effect in the experimental animal than are required for exotoxins. The effects produced by endotoxins on the host are systemic such as fever and general body reactions, rather than strictly neurological effects, as is the case with most exotoxins. Endotoxins are found in the gram-negative bacteria mostly and are obtained subsequent to the death and autolysis of the cells. The endotoxins are extracted from and associated with the cell structure (cell wall). Good examples of pyrogen producing bacteria are S. typhosa, E. coli, and Ps. aeruginosa. Exotoxins are produced during the growth phase of certain kinds of bacteria and are liberated into the medium or tissue. Exotoxins are protein in nature and their reactions are specific. For example, Clostridium botulinum produces an exotoxin of unusual potency which affects only neurological tissue. Other well-known examples of exotoxins are tetanus toxin, Shiga toxin, and diphtheria toxin. Pyrogens are:
- Known to consist biochemically of a lipid-polysaccharide-peptide substance
- Heat stable at the temperature of boiling water
- Demonstrate a low order of immune response
- Produced from persistent gram-negative bacteraemia which could have a 50% mortality rate

Bactericidal procedures such as heating, filtration, or adsorption techniques do not eliminate pyrogens from parenteral solutions. All ingredients must be kept pyrogen-free in the first place. For this assurance, the manufacturer carries out comprehensive pyrogen screening tests on all parenteral drug ingredients and sees to their proper storage prior to use. Ideally, the manufacturer recognises the critical steps in the manufacturing operations that could allow growth of pyrogen producing bacteria and monitors these areas routinely. For example, the water in the holding tanks would be tested for pyrogens and the manufacturer would insist on minimum holding times so that only pyrogen-free water is used. Pyrogen-free water, as "water-for-injection" outlined in the USP, is the heart of the parenteral industry.

12.3. Pyrogen Assay - Limulus Amoebocyte Lysate

Many laboratories conduct pyrogen assays by means of the limulus amoebocyte lysate (LAL) test method. The LAL method is useful especially for screening products that are impractical to test by the rabbit method. Products best tested for endotoxins by LAL techniques are: radiopharmaceuticals, anaesthetics, and many biologicals. Essentially, the LAL method reacts hemolymph (blood) from a horseshoe crab (limulus polyphemus) with an endotoxin to form a gel. The quantity of endotoxin that gels is determined from dilution techniques comparing gel formation of a test sample to that of a reference pyrogen, or from spectrophotometric methods comparing the opacity of gel formation of a test sample to that opacity of a reference pyrogen. The LAL test is considered to be specific for the presence of endotoxins and is at least a hundred times more sensitive than the rabbit test. Even picogram quantities of endotoxins can be shown by the LAL method. Although LAL is a relatively new pyrogen testing method, it has produced a wide variety of polysaccharide derivatives that give positive limulus test results and also show fever activity. It is also a fact that some substances interfere with the LAL test even when pyrogens are present.

Some firms use the LAL test for screening pyrogens in raw materials and follow up with pyrogen testing on the final product by means of the USP rabbit assay. The LAL test for pyrogens in drugs requires an amendment to the NDA on an individual product basis. LAL test reagents are licensed by the Bureau of Biologics. For devices, a firm must have its protocol approved by the Director Bureau of Medical Devices, before it can substitute the LAL assay for the rabbit. What is certain is that pyrogens remain a potential source of danger with the use of parenteral therapy.

12.4. Endotoxins and Depyrogenation

Endotoxins are used to challenge the effectiveness and consistency of depyrogenation tunnels. Endotoxin challenge vials must be processed through a depyrogenation process that must demonstrate a ≥3 log reduction in endotoxin. Typically, endotoxin challenge vials are placed in close proximity to thermocouples. Using this approach, the temperature profile of the position can be obtained during a cycle. Endotoxin challenge testing is often done during SAT and process development.

Endotoxin challenge testing is typically a requirement of validation, however, no commercial product can be used during a depyrogenation tunnel performance qualification using endotoxins as the product would be potentially contaminated with endotoxins. Therefore, performance validation of depyrogenation processes results in the discarding of the vials or ampules. Depyrogenation tunnels generate tremendous amounts of heat and can operate up to temperatures as high as 320°C (depending on design and operational constraints). Air handling is also a key function of the depyrogenation tunnel. Tunnels should not allow non-sterile air from the room into the sterile air inside tunnel zones. This is done through the use of HEPA filters and an overpressure cascade approach of the tunnel compared to the surrounding room or environment. Air flow must be laminar in nature to ensure the tunnel can maintain the correct pressures and temperatures. Most tunnels are divided into two sections:

I. Hot zone (depyrogenation)

II. Cool zone (sterilisation/cooling)

The depyrogenation section typically operates at higher temperatures in excess of 270°C which is the recognised depyrogenation temperature. Depending on the technical specification of the components, set-points of 290°C, 300°C or 320°C can be used. Components move slowly through the depyrogenation stages of tunnels. The "sterilising cooling section" operation mode sterilises the cooling sections. Sterilisation cycle consists of the following steps:

1. Pressure drop
2. Draining heat exchanger
3. Heating up of cooling sections set value of temperature (e.g. 240°C)
4. Sterilisation cooling sections: keeping temperature at recipe set value for recipe set value of time
5. Cooling down without heat exchanger: until temperature reaches < 95°C
6. Cooling down with heat exchanger: until temperature reaches < 25°C

The key requirement of cool zone sterilisation is that the temperature within the zone is maintained at a minimum of 170°C for a period of no less than two hours. This gives a very high degree of assurance that the zone is sterile and suitable for sterile manufacturing operations to occur. In summary, the endotoxin challenge must be sufficient to demonstrate a ≥3 log reduction in endotoxin.

12.5. Biological Indicators for Dry Heat

Biological Indicators (BIs) (most commonly Bacillus atrophaeus) are used to demonstrate the efficacy of cool zone sterilisation in depyrogenation tunnels. Using a known indicator population and D-value, the delivered lethality needed to obtain an SAL of at least 10-6 can be determined. The lethality of a cycle can be calculated using the below equation:

$$\text{Lethality, } F(h) = \Delta T \times \Sigma L$$

$L = 10(t-t_o) / Z$
$Z = 20$ constant
$t_o = 170$ the base temperature (°C)
t = actual temperature (°C)
ΣL = cumulative sum of time
ΔT = time differential (scan time)

12.6. Control of Materials

Items intended for sterilisation or depyrogenation should be prepared and maintained under conditions that will ensure that pre–sterilisation or depyrogenation levels of bioburden, particulate and pyrogen contamination are minimised. Items that will come into contact with sterile dosage forms, filling equipment, containers and closures after sterilisation or depyrogenation in a dry heat oven should be packed for sterilisation in an appropriate clean environment. An appropriate standard would be environmental grade C or D under local protection by HEPA-filtered air.

12.7. Contamination Considerations

Protection of items against contamination before sterilisation or depyrogenation is not generally an issue when washers and tunnels are integrated. However, components should in all cases be received in packaging that minimises contamination risk (e.g., from fibres) and handled in such a way as to minimise contamination Items should be clearly identified and controlled to avoid mix-ups between sterile and non-sterile items. Chemical indicators may be attached to containers or placed within loads.

These indicate, through a colour change, that items have been exposed to steriliser conditions but cannot be taken as proof of the adequacy of sterilisation cycles. However, if chemical indicators do not change colour they should be interpreted as confirming sterilisation failure. Items that are not dried immediately after cleaning should be sterilised as soon as possible (no longer than eight hours and preferably within four hours of cleaning) to minimise the risk of microbial proliferation and eventually pyrogen formation between cleaning and sterilisation.

A maximum storage time before re-sterilisation should be specified in case the equipment is not used immediately. Adequate cleaning, drying, and storage of equipment provide for control of bioburden and prevent contribution of endotoxin load.

12.8. Start-up Conditions

In sterilising ovens, following any drying phase, the load is typically heated up by closing the dampers to the fresh air supply. Air within the oven is continually recirculated over heating elements and through HEPA filters. In modern sterilisers the cycle is usually under automatic control. If the steriliser requires manual intervention for adjustments (e.g. dampers), then this should be very clearly and precisely defined in the operating SOP and details recorded on the record of each sterilisation cycle. In tunnels, the heating occurs as the components progress into the heating zone.

Control and monitoring should be independent and operate from different temperature sensors. Normally, temperature control and routine monitoring is by fixed position chamber sensors. The relationship to load temperature is established by the validation. If there are movable permanently installed temperature sensors, then these should be placed within the oven chamber and within the most difficult to heat position of the load as determined during validation.

This should be very clearly and precisely defined in the operating SOP and details recorded on the record of each sterilisation cycle. Where only data from fixed position chamber sensors are available, the chamber sensor should be positioned in the same position as used in the validation, generally the most difficult to heat position of the chamber. An appropriate allowance for lag phase should be included in the standard cycle (e.g. to set the steriliser timer). This approach is used to compensate for load lag times (the time difference between chamber probes and the load cold point reaching sterilising temperature) as established during validation as part of performance qualification. Note that this correction for lag should be part of the standard cycle as defined by validation and incorporated into the operating SOP, included in the automatic or manual control (as applicable) and included as part of the master process record (mpr) or acceptance criteria used to assess the cycle.

Tunnel control and monitoring should be independent, and operate from different temperature sensors.

The control and routine monitoring of tunnels is by fixed position sensors. The relationship between tunnel and load temperature is established by the validation. The tunnel sensors should be positioned in the same position as used in the validation. The acceptance criteria for the cycle are set on the basis of the validation data such that the tunnel load receives the correct heat input. For tunnels, the sterilisation process is continuous and so the temperature record is of a set temperature. Thus, in order to verify heat treatment of components, the belt speed should be confirmed and, if adjustable, recorded either continuously or intermittently (at least at start and end during each day of operation).

12.9. In-Process Controls

Process parameters that are essential to sterility assurance should be verified and documented for every load processed. Other less critical process parameters that may be indicative of actual or potential steriliser failure should be verified at a lower frequency. Periodic checks:

- Confirmation of instrument calibration and performance of any applicable calibration checks
- Data to be obtained, documented and verified for each cycle
- Identification of the contents of the load
- Confirmation of compliance with validated loading pattern (ovens only)
- Confirmation of correct sealing of doors
- Confirmation of correct differential pressures
- Continuous record of the temperature, time, belt speed where applicable, throughout each cycle from at least one sensor

12.10. Cooling

Oven loads are generally cooled by switching off the heating elements and opening the fresh air dampers, which allows cool HEPA-filtered air to circulate around the load. The rate of cooling should be a compromise between rapidity and the need to avoid product damage. In particular, glass components may be adversely affected by internal stresses caused by rapid and uneven cooling. Note that the cooling phase is established in the validation and fixed as part of the standard cycle.

12.11. Failure of Depyrogenation

The checks on cycle records are vital as a failure of sterilisation or depyrogenation cycles may not be readily detectable in the product testing as there are no visible or practicable non-destructive means of testing for sterility.

The assurance of sterility is thus very heavily based upon the validated process conditions being consistently reproduced during routine operation. It is essential that any failures are promptly detected and that there is a clearly defined course of action in the appropriate operating procedure. Any cycle that does not meet any of its acceptance criteria should be thoroughly investigated.

Materials processed through such a cycle cannot be released solely on passing a test for sterility. Any abnormal or unusual occurrences should be formally recorded on the appropriate site documentation and notified to production management and quality management (even if the occurrence is not formally part of acceptance criteria). They should then be assessed for impact on the sterilisation or depyrogenation and on the functionality of the unit. Procedures should be in place to address such situations (e.g. containment measures). There must be a formal, thorough and fully documented investigation of all cycle failures under the site failure investigation procedure. Possible causes of sterilisation or depyrogenation failure(s) include but are not restricted to:

- Components held for insufficient time during sterilisation
- Too low of a depyrogenation temperature in the hot zone

This may happen in the event of the load lag time being longer than expected due to use of unapproved load patterns, over-loaded ovens, inadequate drying etc.

- Ingress of non-sterile air due to inadequate over-pressure, faulty door seals or filter failure
- Dampers failing to operate correctly
- Excessive residual water in containers (from washing stage)

12.12. Depyrogenation -Performance Qualification (PQ)

Performance Qualification examines the effectiveness and reproducibility of the Depyrogenation cycles in respect of a particular load. Thermocouples are used to verify the set temperatures are reached and maintained throughout the cycle and at various points in the tunnel to demonstrate consistent heat distribution. Endotoxin challenge vials or ampules are also used to demonstrate reduction in endotoxin levels to within acceptable levels and hence ensuring products are safe for patient use. Traditionally, the theme of consistency in Performance Qualifications has been demonstrated by completed 3 distinct batches or runs as part of PQ. Although Risk based approaches to validation (e.g. ASTM E2500 etc.) are increasingly used across the life science industry. Completing a minimum of 3 batches or runs for initial Performance Qualification is still the expected requirement.

The size of the glass components (vials, ampule's or cartridges) must also be considering in the design of Performance Qualifications. The efficacy of the tunnel or sterilising effect is ultimately determined by the size, shape and mass of the components processed. A family or bracketing approach may be utilised to reduce the amount of runs or cycles required. For example, if a manufacturing process utilising 4 different vial sizes across a product – 5ml, 10ml, 15ml and 20ml. Based on technical rationale and some level of evidence, the 5ml and 20ml vials could be considered "worst case". The 5ml been the smallest may exhibit the smallest nick sizes and internal geometry. The 20ml would in this case be the largest vial, with the largest surface area and mass – another worst case configuration.

Another consideration is the position and quantity of thermocouples. For PQ, a rationale based on sound science must support the locations and quantity of thermocouples used during the validation. This should be based on historic data if available, or in the case of new equipment, data generated during SAT testing and/or Engineering development studies.

The thermocouple placement should also be carefully considered. If a particular location is to be assessed, the thermocouple should be secured with Kapton tape. Movement of the probe during a test cycle may result in the data been inconclusive or deemed non-representative.

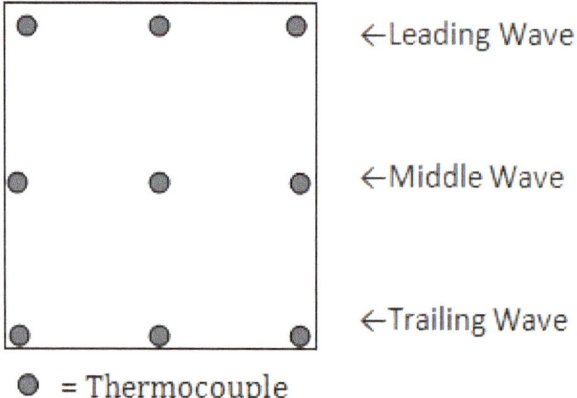

Simple representation of a Depyrogenation tunnel (rectangle). Tunnels are filled with components which progress along a belt. Therefore, each batch shall have a leading wave, middle wave and trailing wave. Place thermocouples at the above locations should verify the temperatures at each location as the product moves through the Depyrogenation tunnel. Endotoxin challenge samples may also be placed in similar locations.

Thermocouples – Points to consider

- The point of contact with the component (e.g. bottom of vial, middle of vial etc.)
- Size of component if a bracketing approach is chosen

Materials processed through such a cycle cannot be released solely on passing a test for sterility. Any abnormal or unusual occurrences should be formally recorded on the appropriate site documentation and notified to production management and quality management (even if the occurrence is not formally part of acceptance criteria). They should then be assessed for impact on the sterilization or depyrogenation and on the functionality of the unit. Procedures should be in place to address such situations (e.g., containment measures). There must be a formal, thorough and fully documented investigation of all cycle failures under the site failure investigation procedure. Possible causes of sterilization or depyrogenation failure(s) include but are not restricted to:

- Components held for insufficient time during sterilization
- Too low Depyrogenation temperature in the Hot Zone

This may happen in the event of the load lag time being longer than expected due to e.g. use of unapproved load patterns, over-loaded ovens, inadequate drying etc.

- Ingress of non-sterile air due to inadequate over-pressure, faulty door seals or filter failure
- Dampers failing to operate correctly
- Excessive residual water in containers (from washing stage)

13. Aseptic Processing

Medical devices and healthcare products are often required to be sterile for use as diagnostic devices, treatment regimes or other clinical applications. It is largely the preference to sterilze products in their final configuration, (terminal sterilization) however, this is not always possible, depending on the product and the manufacturing requirements. For products that cannot be terminally sterilized, aseptic processing offers an alternative to deliver sterile products. This is therefore necessary for many pharmaceutical and biological treatments that require sterility to protect their efficacy and are delivered via injection into the skin, muscle or bloodstream.

Aspetic processing depends on the pre-sterilization of product, components such as primary packaging (e.g. glass vial, elastomeric stopper) and all equipment and utensils coming into contact with the product before they are aseptically processed. The aseptic process then is required to maintain the sterile conditions during manufacturing with the final container of product been sterile. Asepetic processing must control sources of contamination. Any point where materials are introduced or operator intervention is required can lead to microbial contamination that can render the product unsterile and risk patient safety.

Sterile manufacturing operations depends on several factors including the right design and operation of facilities, utilities and equipment. Sterility assurance must be demonstrated to be in control within a manufacturing setting. This is achieved by:
- Qualification and validation of the processes, facilities, utilities, equipment, cleaning methods and sterilisation operations
- Qualified personnel for aseptic handling in conventional clean rooms or by barrier systems
- Control of critical aspects and critical parameters via the application of change management, change control and a suitable quality management system
- Environmental monitoring
- Routine Maintenance
- Analytical method validation

The impact of contaminated injectable products can result in serious illness or death to patients. Many injectable treatments sustain life and bio-chemical processes or genetic conditions. While there is always residual risks or acceptable risks, it is important to mitigate against any risks throughout the manufacturing process. Furthermore, a risk-based approach to operations and in particular changes to the process must be maintained throughout the life cycle of a product. Contamination can be caused by particles or microbes. Where appropriate and technically permissible terminal sterilisation is the preferred point of sterilisation. Terminal sterilisation is when the final sealed product in its container is sterilised at the end of the process.

Isolator systems are used to facilitate aseptic processing and are designed to maintain a sterile environment under certain conditions and controls. The space within the isolater provides a segrated and controls environment that is protected from the operator and the surrounding environment which can contain particulate and micro-organisms. Isolators are equipped with transfer systems to support the introduction of pre-sterilised parts and components into the isolator. Materials been introduced need to be appropriately sterilised and biocontaminated or come in sterile packaging that can be sanitized, cleaned and transferred according to a contollred procedure.

The basic principle of isolation technology is to separate a process from the environment. This can be achieved with two different solutions: a traditional grade b cleanroom with laminar flow cabinet (lfc) or a more flexible solution such as an isolator. Getinge isolators narrow containment to just around the process within the chambers. Our solutions minimize downtime and avoid contamination and false positives during sterility testing of sterile drugs, components, and medical devices. Running costs are relatively low – sometimes as low as 20% of the costs of a cleanroom solution, as only the conditions inside the isolator need to be controlled. Either unidirectional air flow (udaf) or engineered turbulent air flow (etf) can be applied, depending on your process needs or local regulations. In an isolator system, standardized, pre-tested components are combined into a customized overall solution. Most of the technical solutions such as integrated h_2o_2 generator are common across the range. A modular isolator system provides flexible configuration and modification possibilities to meet your specific process and application requirements.

The transfer of material into and out of isolators requires specific technologies. Getinge La Calhène is the originator and manufacturer of the Rapid Transfer Port (RTP) or Alpha-Beta transfer ports. This system is now the industry standard for transfer of aseptic or toxic products in biomedical research institutions and pharmaceutical factories all over the world. Supporting regulatory compliance To ensure safety, all Biopharmaceutical production facilities must comply with strict regulations. These include current Good Manufacturing Practices (cGMP) associated with production of the finished product, and current Good Laboratory Practices (cGLP) associated with quality testing related to the product. The components in an isolator system are pre-tested resulting in reliability at a very high level and facilitated validation of equipment. Each process and cycle must follow validated customer requirements to deliver stable and repeatable results in the shortest time and at a minimum cost. Getinge provides a fully documented on-site validation service to fulfill customer and regulatory requirements.

13.1. Design Considerations for Isolator Systems

The design of the majority of isolator systems is based on using positive pressure conditions as the operating principle. The air supply feeding the isolator is filtered using a high efficiency particulate air filter or ultra high efficiecny if required. The combination of positive pressure and clean air forms the basis of the sterile aseptic enviromonemt. However, decomtamination of the surfaces internally is necessary priort to production or filling operations. This is most commonly achieved using Vaporised hydrogen peroxide. Any items or utensils that are transferred aseptically into the isolator need to be sterilised prior to entry. This can be done via mosit heat sterilisation e.g. scissors, tweezers etc.

13.2. Definition of Aseptic Processing

> Aseptic processing is the combination of unit operations (manufacturing steps) that need to be combined together while maintaining sterility

13.3. Regulations and Standards

Aseptic processing is a requirement per current good manufacturing practice (CGMP) regulations (2l CFR parts 210 and 211) when manufacturing sterile drug and biological products using aseptic processing. Further guidance and requires can be found in ISO 13408, Aseptic processing of health care products, which has the following subparts:

Part 1: General requirements
Part 2: Filtration
Part 3:Lyophilization
Part 4: Clean in place technologies
Part 5: Sterilization in place
Part 6 :Isolator systems

13.4. Technical Comparison of Terminal Sterilization and Aseptic Processing

Previously, the applications of terminal sterilization have been detailed along with their strengths, however, terminal sterilization does not suit all products or processes. Healthcare Products that are subject to terminal sterilization are manufactured in controlled environmental conditions. The environmental controls work to control and minimize the microbial and particulate content to help ensure that the subsequent sterilization process is successful. Therefore during the manufacturing process for products to be processed via terminal sterilization, control of the environment aims to ensure low bioburden conditions are present. For aseptic manufacturing the control of the environment is also required, however to a much higher quality standard (e.g. cleanroom classification). Depending on the product type and technology of the process, a number of processes and manufacturing steps are necessary. For example, a drug product that is filled into a vial may include the following processing steps:

    I.   Formulation and compounding of the drug product

II. Washing, depyrogenation and sterilization of the vial
III. Sterilization of the vial cap or stopper
IV. Aseptic filling
V. Inspection
VI. Secondary Packaging

The above steps I-V must be completed under sterile conditions in order to produce a sterile finished product as there is no process that can sterilize the product in its final container (vial). Prior to the introduction of the drug product and components to the manufacturing process, certain equipment, processing area's and utensils must be sterilized.

I. Formulation and compounding of the drug product
   a. liquid dosage forms are often subjected to filtration
II. Washing, depyrogenation and sterilization of the vial
   a. Washing of vials or containers. Washing removes any particles or other foreign matter. Water for sterilization, ultrasonics and air pressure can be used in the washing process.
   b. Depyrogenation glass containers are subjected to dry heat to remove pyrogens from the vial if present.
   c. Sterilization of the vials are required prior to filling and capping/closure
III. Sterilization of the vial cap or stopper
   a. rubber closures or caps are subjected to moist heat to sterilize components.
IV. Aseptic filling
   a. Any manual or mechanical manipulation of the sterilized drug, components, containers, or closures prior to or during aseptic assembly poses the risk of contamination. The filling area should exist in an isolater and be deemed sterile and decontaminated prior to use.
V. Inspection
   a. At this point the product is protected from gross contamination from the environment. However, controls should ensure that the inspection area is fit for purpose and it maintained to the specified clean room/controlled conditions.
VI. Secondary Packaging
   a. This may involve assembly of vials or containers into deliver pens or devices, or bulk transportation to another site for further assembly and packaging.

Principles
- The Application of CGMP is essential for the manufacturing facility
- Each manufacturing processes requires validation and control
- Each process has the potential to introduce an error that may lead to the distribution of a contaminated product
- Aseptic processing requires that sterility is maintained through all the stages and unit operations including preparation, manufacturing, filling and sealing vials or containers.
- Contamination control is critical at each step, if microbial contamination is introduced to the product, it cannot be eliminated in the same manner if terminal sterilization is used.
- The equipment for aseptic production should be designed and validated in order to confirm it is fit for its intended use.

13.5. Isolator and Glove Access

Access to the sterile environment within an isolator is facilitated by Glove access for operators or technicians. While the access points are often necessary, they also present a risk of leakage and contamination of the sterile zone. For example, a faulty//damaged glove or sleeve (gauntlet) provides a mechanism for contamination to occur. Glove leak testing or glove integrity testing is necessary in order to test and ensure no ingress from the outside- nonsterile area can occur. Selection of the right glove design, replacement frequencies, leak and integrity test methodology, visual inspection etc. need to be addressed by responsible functions. The choice of durable glove materials, coupled with a well-justified replacement frequency, are key aspects of good manufacturing practice to be addressed. If manufacturing operations are ongoing and a leak or fault with a glove system arises, it can impact the product and cause critical defects and risks contamination with the sterility of the product compromised. Product may therefore need to be discarded which can be costly and requires additional resources and labor of personnel.

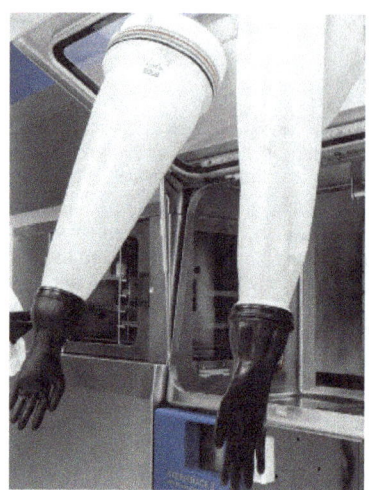

13.6. Isolator Design Requirements

There are two types of aseptic processing isolators: open and closed. Closed isolators employ connections with auxiliary equipment for material transfer. Open isolators have openings to the surrounding environment that are carefully engineered to segregate the inner isolator environment from the surrounding room via overpressure.

Turbulent flow can be acceptable within closed isolators, which are normally compact in size and do not house processing lines.

Other aseptic processing isolators employ unidirectional airflow that sweeps over and away from exposed sterile materials, avoiding any turbulence or stagnant airflow in the area of exposed sterilized materials, product, and container closures. In most sound designs, air showers over the critical area once and then is systematically exhausted from the enclosure. The air handling system should be capable of maintaining the requisite environmental conditions within the isolator.

13.7. Materials of Construction

As in any aseptic processing design, suitable materials should be chosen based on durability, as well as ease of cleaning and decontamination. For example, rigid wall construction incorporating stainless steel and glass materials is widely used. Pressure Differential Isolators that include an open portal should be designed to ensure complete physical separation from the external environment

. A positive air pressure differential adequate to achieve this, separation should be employed and supported by qualification studies. Positive air pressure differentials from the isolator to the surrounding environment have largely ranged from approximately 17.5 to 50 Pascals.

The appropriate minimum pressure differential established by a firm will depend on the system's design and, when applicable, its exit port. Air balance between the isolator and other direct interfaces (e.g., washer and depyrogenation tunnel /dry heat tunnel) should also be qualified.

The positive pressure differential should be coupled with an appropriately designed opening to the external environment to prevent potential ingress of surrounding room air by induction. Induction can result from local turbulent flow causing air swirls or pressure waves that might push extraneous particles into the isolator.

Local Class 100 (ISO 5) protection at an opening is an example of a design provision that can provide a further barrier to the external environment.

Clean Area Classifications The interior of the isolator should meet Class 100 (ISO 5) standards.

The classification of the environment surrounding the isolator should be based on the design of its interfaces (e.g., transfer ports), as well as the number of transfers into and out of the isolator. A Class 100,000 (ISO 8) background is commonly used based on consideration of isolator design and manufacturing situations. An aseptic processing isolator should not be located in an unclassified room.

13.8. Isolator Access

The ability to maintain integrity of a decontaminated isolator can be affected impacted by the design of transfer ports. Various adaptations, of differing capabilities, allow for the transfer of supplies into and out of the isolator. Multiple material transfers are generally made during the processing of a batch.

Frequently, transfers are performed via direct interface with manufacturing equipment.
Properly maintained and operated rapid transfer ports (RTPs) are an effective transfer mechanism for aseptic transfer of materials into and out of isolators.

Some transfer ports might have significant limitations, including marginal decontaminating capability (e.g., ultraviolet) or a design that has the potential to compromise isolation by allowing ingress of air from the surrounding room. In the latter case, localized HEPA-filtered unidirectional airflow cover in the area of such a port should be implemented.

Isolators often include a mousehole or other exit port through which product is discharged, opening the isolator to the outside environment. Sufficient overpressure should be supplied and monitored on a continuous basis at this location to ensure that isolation is maintained.

13.9. Isolator Decontamination

Decontamination of the surface

Surface Exposure Decontamination procedures should ensure full exposure of all isolator surfaces to the chemical agent. The capability of a decontaminant to penetrate obstructed or covered surfaces is limited. For example, to facilitate contact with the decontaminant, the glove apparatus should be fully extended with glove fingers separated during the decontamination cycle. It is also important to clean the interior of the isolator per appropriate procedures to allow for a robust decontamination process

Efficacy

The decontamination method should render the inner surfaces of the isolator free of viable microorganisms. Multiple available vaporized agents are suitable for achieving decontamination. Process development and validation studies should include a thorough determination of cycle capability

The characteristics of these agents generally preclude the reliable use of statistical methods (e.g., fraction negative) to determine process lethality An appropriate, quantified Biological Indicator (BI) challenge should be placed on various materials and in many locations throughout the isolator, including difficult to reach areas.

Cycles should be developed with an appropriate margin of extra kill to provide confidence in robustness of the decontamination processes. Normally, a four- to six-log reduction can be justified depending on the application.

The specific BI spore titer used and the selection of BI placement sites should be justified. For example, demonstration of a four-log reduction should be sufficient for controlled, very low bioburden materials introduced into a transfer isolator, including wrapped sterile supplies that are briefly exposed to the surrounding cleanroom environment.

The uniform distribution of a defined concentration of decontaminating agent should also be evaluated as part of these studies
Chemical indicators may also be useful as a qualitative tool to show that the decontaminating agent reached a given location.

Frequency

The design of the interior and content of an isolator should provide for its frequent decontamination. When an isolator is used for multiple days between decontamination cycles, the frequency adopted should be justified. This frequency, established during validation studies, should be reevaluated and increased if production data indicate deterioration of the microbiological quality of the isolator environment.

A breach of isolator integrity should normally lead to a decontamination cycle. Integrity can be affected by power failures, valve failure, inadequate overpressure, holes in gloves and seams, or other leaks. Breaches of integrity should be investigated. If it is determined that the environment may have been compromised, any product potentially impacted by the breach should be rejected.

Filling Line Sterilization

To ensure sterility of product contact surfaces from the start of each operation, the entire path of the sterile processing stream should be sterilized. In addition, aseptic processing equipment or ancillary supplies to be used within the isolator should be chosen based on their ability to withstand steam sterilization (or equivalent method). It is expected that materials that permit heat sterilization (e.g., SIP) will be rendered sterile by such methods. Where decontamination methods are used to render certain product contact surfaces free of viable organisms, a minimum of a six-log reduction should be demonstrated using a suitable biological indicator.

Environmental Monitoring

An environmental monitoring program should be established that routinely ensures acceptable microbiological quality of air, surfaces, and gloves (or half-suits) as well as particle levels, within the isolator. Nutrient media should be cleaned off of surfaces following a contact plate sample. Air quality should be monitored periodically during each shift. For example, we recommend monitoring the exit port for particles to detect any unusual results. Media used for environmental monitoring should not be exposed to decontamination cycle residues, as recovery of microorganisms would be inhibited.

Rubber closures (e.g., stoppers and syringe plungers) can be cleaned by multiple cycles of washing and rinsing prior to final steam or irradiation sterilization. At minimum, the initial rinses for the washing process should employ at least Purified Water, USP, of minimal endotoxin content, followed by final rinse(s) with WFI for parenteral products.

Normally, depyrogenation can be achieved by multiple rinses of hot WFI. The time between washing, drying (where appropriate), and sterilizing should be minimized because residual moisture on the stoppers can support microbial growth and the generation of endotoxins. Because rubber is a poor conductor of heat, extra attention is indicated in the validation of processes that use heat with respect to its penetration into the rubber stopper load Validation data from the washing procedure should demonstrate successful endotoxin removal from rubber materials.

A potential source of contamination is the siliconization of rubber stoppers. Silicone used in the preparation of rubber stoppers should meet appropriate quality control criteria and not have an adverse effect on the safety, quality, or purity of the drug product. Contract facilities that perform sterilization and/or depyrogenation of containers and closures are subject to the same CGMP requirements as those established for in-house processing.

The finished dosage form manufacturer should review and assess the contractor's validation protocol and final validation report. In accord with 211.84(d)(3), a manufacturer who establishes the reliability of the supplier's test results at appropriate intervals may accept containers or closures based on visual identification and Certificate of Analysis review.

13.10. Isolator Barrier Systems

Equipment referred to as seperation systems are necessary in providing protection and contamination control in the critical processing areas of an acceptance processing area. The need to prven contamination from both non viable particule and microbiological contamination. Seperation systems include controlled air flow devices that use the principles of air flow and barriers to control the envirnonment.An isolator (barrier system) is also a speration system but is also designed to support aseptic processing and manufacturing by allowing operator intervention into the controlled envrionment via glove-sleeve systems. The supplied air to such systems is generally supplied through a microbially-retentive filtration system. High efficiency particulate air (HEPA) filters are capable of removing particles as small as 0.3μm making them an integral part of isolator technology. HEPA filters should be capable of achieving Grade A (ISO Class 4.8) at-rest and in-operation. Some exceptions are permitted, such as powder filling, however, risk assessments should mitigate risk to patients. The isolator is a sealed enclosure where there is no direct opening to the external environment or room. Transfer of materials or utensils is done in a controlled manner using a decontaminated interface. Movement of materials and items in and out of isolators presents the most common risks and failure modes in introduction contamination. Any operator intervention during processing must follow strict requirements and follow aseptic technieque.

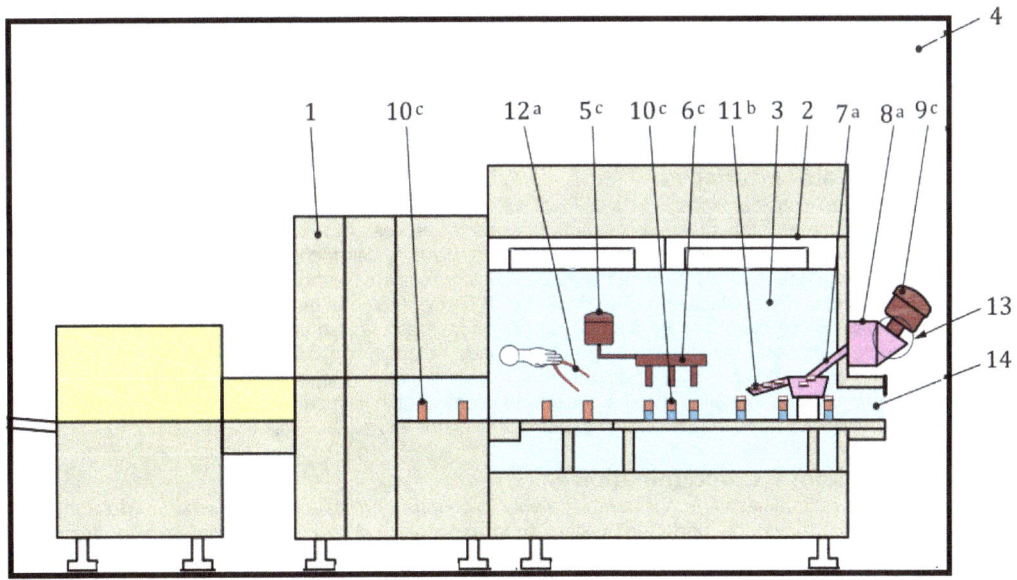

Key

1	Depyrogenation tunnels	11	Stopper
2	Isolator	12	Forceps
3	Controlled environment	13	RTP type transfer port
4	Surrounding environment	14	Mouse hole
5	Vessel for filling liquid		
6	Filling nozzle and tubing		
7	Stopper bowl and shoot (located inside of isolator)	a	Forceps or other utensils are product contact surfaces.
8	Receiving vessel for stopper (include the alpha port and located outside of isolator)	b	Only the surface that can come into contact with product.
		c	Inside surface is critical.
9	Mobile vessel for stopper (include the beta port)		
10	Vial		

13.11. Isolator Interfaces

Depending on the design considerations and individual vendor designs, isolators can have a number of operation interfaces. The term "interface" refers to the ability of an operator or process technician to interact with the machine. The primary method of intervention utilizes glove systems. Four part glove systems consisting of a gauntlet, glove, cuff-ring and sleeve. When used properly and by trained personnel, glove systems support critical line interventions required during aseptic processing and manufacturing. The gasket of RTP systems has been identified as a potential source of contamination in isolators since there may be a small contact surface around the gasket that may not be exposed to the decontaminating agent. A risk analysis should be done to evaluate potential contamination risks with the gaskets and the need for maintenance programmes. Transfer of material into and out of the isolator is also a potential source of contamination. Furthermore, isolators may also be designed in combination with smaller enclosures associated with them to allow the continuous ingress of materials through the smaller isolator into a main isolator.

13.12. Isolator Decontamination

The purpose of bio-decontamination is to remove viable bioburden on exposed surfaces inside the isolator; a decontamination process should be performed using sporicidal chemical agents associated with decontamination equipment such as gas/vapour phase decontamination systems using hydrogen peroxide (e.g. VHP) or the equivalent. A decontamination cycle is an automated machine cycle that is controlled and monitored during each stage of the cycle. Cycles can be divided into four stages:

I. -Dehumidification
II. -Conditioning
III. -Decontamination
IV. -Aeration

Dehumidification: The dehumidification stage (also known as pre-conditioning) is designed to ensure that the isolator enclosure has a predefined humidity value (< 20 % RH) to ensure a proper concentration of decontaminating agent.

Conditioning: Depending on the complexity of the system, at a minimum, the isolator must have a tightly controlled temperature range, positive pressure and air velocity control. During this initial stage, the isolator doors and ports must be closed and sealed. Any defects in the barrier system should result in an alarm and abort the cycle. During conditioning, an automated leak test should be initiated to detect any breaks in the barrier system (e.g. defective gloves or seals). Heating of VHP delivery pipework also occurs. The conditioning stage is when the decontaminating agent shall reach the minimum concentration required to achieve the desired microbial reduction.

Decontamination: At this stage the VHP is maintained in the isolator according to the dosing rate contained in the recipe or cycle settings. The time and total amount of VHP must result in a kill in BIs placed within the isolator. Generally a 6 log reduction is required for a cycle to be deemed a success.

Aeration: During the aeration stage the amount of residual decontaminating agent must fall to safe levels. (< 1ppm). This is done by blowing the hydrogen peroxide carrying air out of the barrier system using fresh air.

Recommended Critical Process Parameters

I. Amount of H_2O_2 during conditioning (g)
II. Dosing rate (conditioning) (g/min)
III. Time for conditioning (mins)
IV. Amount of H_2O_2 during decontamination (g)
V. Dosing rate decontamination (g/min)
VI. Time for decontamination (mins)
VII. Aeration time (mins)
VIII. Decontamination Agents

Decontamination of isolators is achieved by the supply of gaseous sporicidal agents. These agents must be capable of killing both bacterial endospores and fungal spores. The system typically turns liquid agents into a gaseous vapour. The decontamination agent typically used in industry is hydrogen peroxide. Other agents include formaldehyde, peracetic acid and chlorine dioxide. The rationale for selecting a particular agent should be based on technical data, sporicidal efficacy and the materials and products that come into contact with such agents. Often the starting point when selecting an agent is the manufacturer's recommendations.

The below factors should be considered with regards to biodecontamination:

- Ensure as much surface area as possible of components are exposed.
- Minimise loads in order to limit the bioburden levels prior to the cycle starting.
- For filling and closing machines, design automation to ensure parts are moving during the cycle to facilitate exposure to the agent.
- Ensure all areas are dry and free of foreign objects and debris.

13.13. Facility Layout for Aseptic Processing

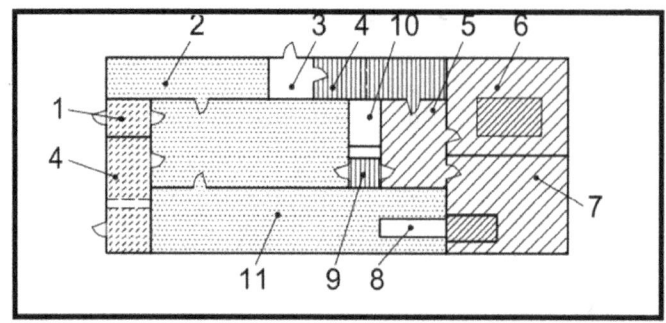

 critical processing zone
 direct support zone
 direct support zone at rest
 indirect support zone
 indirect support zone at rest
 non-classified areas

1 Material Airlock for Materials transfer into area
2 Preparation room
3 Airlock for personnel
4 Changing Area
5 Cooling area
6 Aseptic Area
7 Aseptic Area
8 Vial Washer and Tunnel
9 Material Airlock
10 Steam Sterilizer
11 Final Washing

13.14. Air Classifications

Clean Area Classification (0.5 um particles/ft^3)	ISO Designation[b]	≥ 0.5 μm particles/m^3	Microbiological Active Air Action Levels[c] (cfu/m^3)	Microbiological Settling Plates Action Levels[c,d] (diam. 90mm; cfu/4 hours)
100	5	3,520	1[e]	1[e]
1000	6	35,200	7	3
10,000	7	352,000	10	5
100,000	8	3,520,000	100	50

a- All classifications based on data measured in the vicinity of exposed materials/articles during periods of activity.
b- ISO 14644-1 designations provide uniform particle concentration values for cleanrooms in multiple industries. An ISO 5 particle concentration is equal to Class 100 and approximately equals EU Grade A.
c- Values represent recommended levels of environmental quality. You may find it appropriate to establish alternate microbiological action levels due to the nature of the operation or method of analysis.
d- The additional use of settling plates is optional.
e- Samples from Class 100 (ISO 5) environments should normally yield no microbiological contaminants.

13.15.   Filling Operations

Suspensions and solutions that are filled in glassware such as vials provide lifesaving and sustaining medical treatments for millions of patients worldwide. When the product reaches the filling unit operation, it has been through many unit operations. The product and components must be sterile at this point. Transfer of product to individual vials or containers may be facilitated by employing piston valves, pressure control and peristaltic pumps.

Once the required quantity of solution or suspension has been filled, the next unit operation required is container closure achieved by the insertion or application of a stopper or cap. Key consideration for filling and closing operations include:
  I.   Design and function of filler heads
  II.   Design and function of filler needles
  III. Fill accuracy and fill weight

The filling of Biotechnology Derived Products (BDP) into ampules or glass vials presents similar problems as with the processing of conventional products. Attempting to develop a site, prove clinical effectiveness and safety, as well as the validation of sterile operations, equipment, processes and systems often necessitates a lengthy process to achieve success for a start-up BDP facility.

The batch size initially produced by a BDP is likely to be small. Because of the small batch size, filling lines may not be as automated as for other products typically filled in larger quantities. Thus, there is more involvement of people filling these products. This can present more chances of contamination meaning any operation or involvement must be controlled and monitored.

Problems that have been identified during filling include inadequate attire, deficient environmental monitoring programmes, hand-stoppering of vials, particularly those that are to be lyophilised and failure to validate some of the basic sterilisation processes. Because of the active involvement of people in filling and aseptic manipulations, the number of persons involved in these operations should be minimised, and an environmental programme should include an evaluation of microbiological samples taken from people working in aseptic processing areas.

Another concern about product stability is the use of inert gas to displace oxygen during both the processing and filling of the solution. As with other products that may be sensitive to oxidation, limits for dissolved oxygen levels for the solution should be established. Likewise, validation of the filling operation should include parameters such as line speed and location of filling syringes with respect to closure, to ensure minimal exposure to air (oxygen) for oxygen-sensitive products. In the absence of inert gas displacement, the manufacturer should be able to demonstrate that the product is not affected by oxygen.

Typically, vials to be lyophilised are partially stoppered by machine. However, some filling lines have been observed that utilise an operator to place each stopper on top of the vial by hand. The concern is the immediate avenue of contamination offered by the operator. The observation of operators and active review of filling operations should be performed. Another major concern with the filling operation of a lyophilised product is assurance of fill volumes. A low fill would represent a sub-potency in the vial. Unlike a powder or liquid fill, a low fill would not be readily apparent after lyophilisation, particularly for a product where the active ingredient may be only a milligram. Because of the clinical significance, sub-potency in a vial can potentially be a very serious situation. A common method of filling vials consists of a two-step filling process. Generally, the first step fills up to 90% of the vial, with the second more accurately filling the remaining amount. The following parameters must be maintained to achieve the same fill volumes at each filling cycle:

- Viscosity of the product
- Product temperature
- Pressure in the dosing vessel
- Level in the dosing vessel
- Needle/filling head properties
- Properties of the hose material

13.16. Aseptic Process Simulation

Validation of aseptic processing for products must include simulating the process using aseptic process simulations. For simulations of final product filling, the number of containers filled should be representative of the projected batch size and be sufficient to enable a valid evaluation, including all routine operator interventions.

14. Sterile Barrier Packaging Systems

14.1. Introduction

Sterile barrier systems are packaging solutions that are designed to meet a number of needs. Medical devices, depending on the intended use and their classification often need to be provided to the user or patient in a sterile condition. However, before the product is used, there is a need to ensure the products can be stored and distributed supported by appropriate protective packaging. The adoption of sterile barrier systems for packaging medical devices provide an effective way of allowing a manufacturer to package the medical device, sterilize the device and ensure its remains sterile until the delivery and point of use.

One of the most common types of sterile barrier systems in use for medical devices uses a rigid tray (aka. Blister, tub) with a die cut lid. The tray or tub is normally produced separately by blow moulding or injection moulding. The die cut lid is then sealed or bonded to the tray via a heat sealing process or ultrasonic welding.

Another type of sterile barrier system is the use of a flexible pouch with one side consisting of a polymer film and the other side paper. There is a great number of flexible peel pouches available to accommodate the various sizes of medical devices.

Criticality of sterile barriers is recognised by regulators globally and to achieve the design and manufacture of quality products, there is an important role for the implementation of industry standards

ISO 11607-1 and ISO 11607-2 are standards that cover the various stages of the lifecycle of sterile barrier systems.

Design and use of sterile barrier systems in a manufacturing setting should be understood in the context of a CGMP environment. Compliance to Quality, good documentation, good science and the use of industry standards. The general elements required to position sterile barrier systems for success include:

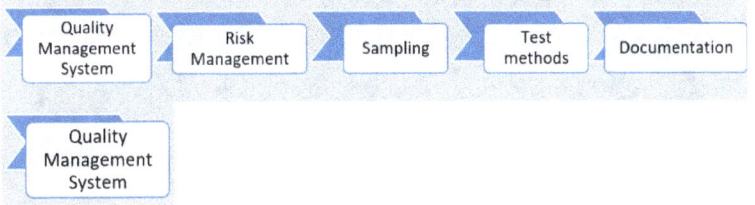

Quality Management System

Quality management systems (QMS) are designed to provide a framework where the output from one internal process forms the input to other processes within the Quality management system. For example, customer complaints relating to packaging can allow manufacturing to respond by making improvements to equipment and processing. Compliance to a QMS ensures that CGMP is practiced which provides the building blocks for designing and manufacturing medical devices. With the application of CGMP via a QMS, sterile barrier systems for medical devices have a strong foundation to support the delivery of quality barrier systems that meet quality and regulatory requirements.

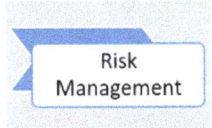

A Risk Management process is a requirement of most Quality Management systems. The Risk management system exhibits its own procedures, requirements and is supported by trained and competent human resources. The purpose of the system is to provide a systematic approach to risk and risk control. Risks relating to sterilization and sterile barriers need to be analyzed, evaluated, controlled and monitored.

Identify hazards & hazardous situations

Design deficiencies in sterile barrier packaging may result in potential hazard

Process failure modes related to sealing, barrier formation can lead to risk of infection

Design or manufacturing issues may present usability issues. e.g. difficulty opening package

Estimate and evaluate risks

Is there a low, medium or high probability of the risk occurring

Determine the severity of the harm (e.g. infection, death, other)

Probability and severity make up the risk evaluation

Control Risks

High risks and unacceptable hazards require risk reduction and risk control measures

Control measures can include manufacturing controls e.g. automated inspection of blister seal.

Monitor Effectiveness of controls

Complaints or adverse events can be trended by reason or category

Is there data identifying packaging defects from the field

The application of risk management includes the following elements:

- Scope of planned risk management activities
- Critieria for risk acceptability
- Activiteis for verification of implementation and effectiveness of control measures

Risk Acceptability
A policy or criteria for determining risk acceptability. The policy provides instruction on how to establish the criteria for acceptability of the overall residual risk or it may alternatively provide the risk acceptability criteria upfront.

Criteria for risk acceptability
Criteria for risk acceptability should be established in advance of any risk management activity or execution of the risk management plan so that guidance is available in determining acceptable risk. The policy is normally is included in a risk management procedures or other quality document.

Evaluation of overall residual risk and acceptability
An evaluation of the overall residual risk and acceptability must be completed in accordance with the risk policy.

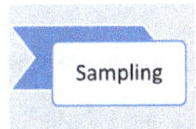

The purpose of sampling and sampling plans is two fold. (1) Sampling of incoming materials that are used in sterile barrier systems is crtical to ensure the input materials are suitable and are in compliance with the material specification required to provide safety and performance. (2) Finished products that utilise sterile barrier systems must be inspected to ensure the out going product quality of lots or batches meets quality standards. Sampling provides confidence that the product meets the quality specifications. Sampling must be statistically based with appropriate sample numbers. Statistical sampling plans can be derived by inhouse expertise and statisicans, or more commonly, the application of standards can provide a statistical basis for sampling.

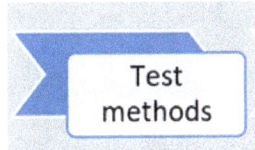

Test methods are the means in which the effective and performance of the packaging system can be tested after manufacturing and at other appropriate times such as during stability. Having a test method documented ensures a consistent approach to testing which gives confidence in the data and results generated. Acceptance criteria should also be detailed along with the rationale for the pass/fail criteria. Test methods for sterile barrier packaging systems include seal integrity testing (dye penetration testing), peel force testing, visual inspection of seal, dimensional inspection of seal area.

Documentation is a requirement of CGMP. In particular, the manufacturing shall have to demonstrate that:
- Records and appropriate documentation is available such as cycle data, test date, validation reports, validation of test methods and assessments to demonstrate conformity with requires regulations and standards.
- Rentetion of data releting to product release, evidence of sterilization, taking into account the expiry of the medical devices and traceability of materials
- Electronic records meet the requirements of relevent regulatory bodies
- Data integrity is evident for all data generated in support of quality and product safety.

14.2. Material Compatability

Materials used to package sterile devices must be suitable as a sterile barrier but also suitable to allow sterilization processing under specified parameters and limits. Variation in materials should also be evaluated to ensure the materials perform as required after sterilization.

14.3. Facators in Design and Development

Product design and development is necessary to comply with regulations, for example 21 CFR Part 820.30, Design Controls. It is also specified across other regional regulatory legislation and ISO 13485. To meet these requirements, procedures for the development of packaging systems must be established, documented, implemented and maintained. For Packaging systems and Sterile Barriers the following factor should be considered:

- Design of packaging is such that it minimizes safety risks to the user and/or patient during intended specified conditions of use.
- Aseptic opening is supported by the design and usability of the package
- Protection from environmental and physical forces so that the medical device remains undamaged and in a sterile condition
- Packaging can be sterilized according to the selected process and it maintains its protection until the product expiry.

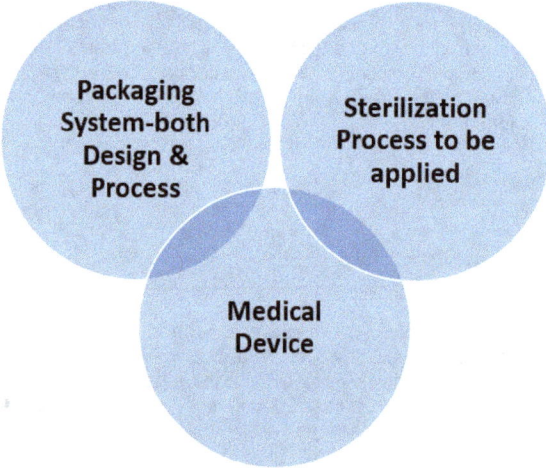

14.4. Performance of Packaging System

Once a medical device is packaged, it needs to remain safe and effective and perform as intended during normal conditions of use. Its performance must be maintained while it is in storage and during distribution. Also, performance must be maintained over the course of the products shelf life and up to the expiry date. Hazards during shipping, storage and distribution can include:

Shock, dropping and vibration
Compression
Temperature
Pressure
Humidity
Transportation

Therefore, performance testing of the sterile barrier system should take into account the above stresses and hazards with testing demonstrating the packaging provides the necessary protection to the medical device.

14.5. Stability of Packaging

Realtime aging must be performed on sterile medical devices to demonstrate the sterile barrier remains effective and intact over the shelf life of the device. Stability testing using Accelerated aging is permitted until data from real time date is available.

4.4.2.2 Guidance on risk management requirements (4.2)

Risks to the user and patient, defined as the combination of the probability of occurrence of harm and the severity of that harm, are by nature inherent to medical devices and related activities and need to be minimized.

Risk management requirements are applicable to all stages of the life cycle of a terminally sterilized medical device, its accessories, and packaging related activities. ISO 11607-1:2019 and ISO 11607-2:2019 cover specifically the packaging life cycle phases of packaging design and development, validation and production.

An ongoing risk management process should be established, implemented, documented and maintained to minimize the risk for the user and the patient. This process should include:

1) identification of hazards and hazardous situations associated with the packaging system,

2) risk estimation and risk evaluation against defined criteria for risk acceptability,

3) risk control,

4) monitoring the effectiveness of the risk control measures.

NOTE 1   Local regulatory requirements can provide mandatory criteria for risk acceptability or these criteria can be based on the generally accepted state of the art.

NOTE 2   PFMEA (Process Failure Mode and Effect Analysis) is an example of a risk analysis tool that is used widely in the industry.

14.6. Lifecycle approach to Sterile Barrier Systems

Design & Development	Validation	Production
Packaging Design per 11607-1 Sealing and assembly process development per 11607-2	Performance and Stability Testing per 11607-1 Usability Evaluation per 11607-1 Process Validation	Process Control Control of Packaging system changes Process Changes Revalidation

14.7. Factors in Sterile Barrier Validation

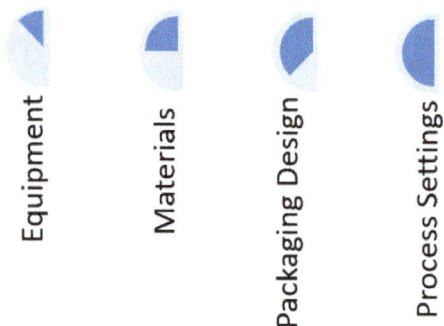

ISO 11607-2: Packaging for terminally sterilized medical devices — Part 2: Validation requirements for forming, sealing and assembly processes

Equipment Validation is necessary for Sterile barrier forming equipment and operation. General principles of CGMP and Validation should apply to the qualification of equipment. Installation Qualification assures that equipment is installed, located and sited in a safe manner that meets the equipment manufacturers requirements.

Operational Qualification of sterile barrier packaging is critical and depends of several physical verifications that provide confidence that the equipment, packaging design, materials and process settings meet the requirements of sterile barriers for medical devices.

Operational Qualification (High) should challenge sterile barrier formation at the high or upper process settings such as (1) seal head temperature, (2) dwell or seal time, (3) force/pressure applied during sealing. These high process settings should be documented in a specification and validation confirms that the high settings can successfully create the sterile barrier and meet acceptance criteria.

Operational Qualification (Low) is when process settings and set to the lower limits of the process. This creates another challenge condition that provides confidence that any minor variations in process settings will not impact on the integrity of the packaging or the sterile barrier system.

OQ testing and verifications
Cosmetic and visual acceptance criteria should be established prior to OQ testing and documented in a specification. Cosmetic issues with sealing or the sterile barrier may be just visual but also may impact seal integrity
Critical dimensions should meet pre-defined acceptance criteria. Drawings should identify critical dimensions and tolerances.
Seal strength is a measure of the sterile barriers seal integrity. A properly formed seal should no voids, bubbles, punctures and so on. Seal width can be a critical parameter depending on the sealing process. An allowable level of minor defects may be permissible depending on the quality specifications. Note: ASTM F88 provides methodology for seal strength measurement.

Performance Qualification (PQ)
Performance Qualification should challenge sterile barrier formation at nominal settings (e.g. dwell time, temperature, pressure). PQ aims to create the processing conditions that will typically be applied during normal production. Therefore, the equipment operation, materials, contents of packaging, operating conditions should be controlled and in accordance with specifications and procedures.

The standards, ISO 11607-2 Packaging for terminally sterilized medical devices — Part 2: Validation requirements for forming, sealing and assembly processes, is often applied by manufacturers and is a requirement of many regulations bodies.
Note: Repeat all tests from OQ in the PQ also.
samples produced at both the upper and lower process limits.

Sterile Pouch packaging

Blister tub with foil (aluminum) sterile barrier lid

Rigid tray containing medical device sealed with die cut lid

Industry Standard
ISO 11607-1, Packaging for terminally sterilized medical devices. Requirements for materials, sterile barrier systems and packaging systems
ISO 11607-2, Packaging for terminally sterilized medical devices. Validation requirements for forming, sealing and assembly processes
EN 868-, Packaging materials for terminally sterilized medical devices. Part 2: Sterilization wrap - Requirements and test methods
EN 868-3, Packaging materials for terminally sterilized medical devices. Part 3: Paper for use in the manufacture of paper bags (specified in EN 868-4) and in the manufacture of pouches and reels (specified in EN 868- 5) - Requirements and test methods
EN 868-4, Packaging materials for terminally sterilized medical devices. Part 4: Paper bags - Requirements and test methods
EN 868-5, Packaging materials for terminally sterilized medical devices. Part 5: Sealable pouches and reels of porous materials and plastic film construction - Requirements and test methods

EN 868-6, Packaging for terminally sterilized medical devices - Part 6: Paper for low temperature sterilization processes - Requirements and test methods
EN 868-7, Packaging for terminally sterilized medical devices - Part 7: Adhesive coated paper for low temperature sterilization processes - Requirements and test methods
EN 868-8, Packaging materials for terminally sterilized medical devices. Part 8: Re-usable sterilization containers for steam sterilizers conforming to EN 285 -Requirements and test methods
EN 868-9, Packaging materials for terminally sterilized medical devices. Part 9: Uncoated nonwoven materials of polyolefines for use in the manufacture of sealable pouches, reels and lids - Requirements and test methods
EN 868-10, Packaging materials for terminally sterilized medical devices. Part 10: Adhesive coated nonwoven materials of polyolefines for use in the manufacture of sealable pouches, reels and lids - Requirements and test methods
ASTM F88, standard test method for seal strength of flexible barrier materials
ASTM F1929, standard test method for detecting seal leaks in porous medical packaging by dye penetration
ASTM F2096, standard test method for detecting gross leaks in packaging by internal pressurization (bubble test)

www.ingramcontent.com/pod-product-compliance
Lightning Source LLC
Chambersburg PA
CBHW062109220526
45471CB00010B/3667